EU Agricultural Law

EUROPEAN MONOGRAPHS

In this series European Monographs this book, *EU Agricultural Law* is the eighty-third title. The titles published in this series are listed at the end of this volume.

EU Agricultural Law

Jens Hartig Danielsen

Wolters Kluwer
Law & Business

Published by:
Kluwer Law International
PO Box 316
2400 AH Alphen aan den Rijn
The Netherlands
Website: www.kluwerlaw.com

Sold and distributed in North, Central and South America by:
Aspen Publishers, Inc.
7201 McKinney Circle
Frederick, MD 21704
United States of America
Email: customer.service@aspenpublishers.com

Sold and distributed in all other countries by:
Turpin Distribution Services Ltd
Stratton Business Park
Pegasus Drive, Biggleswade
Bedfordshire SG18 8TQ
United Kingdom
Email: kluwerlaw@turpin-distribution.com

Printed on acid-free paper.

ISBN 978-90-411-3280-2

© 2013 Kluwer Law International BV, The Netherlands

Printed and Bound by CPI Group (UK) Ltd, Croydon, CR0 4YY.

For Lene, Signe and Anders

Table of Contents

Preface

The European Union's rules on agriculture have played a significant role in the development of Union law in general. If one reviews the most important decisions of the Court of Justice, one will soon see that a large proportion of the cases have been based on the rules of agricultural law. Today, with its comprehensive regulation of issues that are of prime economic importance, Union agricultural law is a rich source of legal disputes. Furthermore, no other area of law is so comprehensively governed by Union rules. Many of these rules have direct effect in national law and directly influence the legal rights of citizens and undertakings. Another equally important aspect of Union agricultural law is its indirect effect, as Union directives in particular are implemented via national legislation. All this means that Union agricultural law cannot be treated as an isolated discipline, set apart from the rest of Union law on its own in a corner. Throughout this book it is made clear that Union agricultural law, as a specialist discipline, must be seen in the context of general Union law.

At the same time, Union agricultural law regulate matters that are of widespread importance to society. In particular, the considerations that lie behind the formulation of agricultural support schemes and the conditions attached to them are of great importance. Union agricultural law deals with problems that have existed for as long as there has been an agricultural support scheme anywhere in the world. In his book *Catch-22*, Joseph Heller gave a somewhat questionable (American) preacher the following words: 'The Lord gave us good farmers two strong hands so that we could take as much as we could grab with both of them... If the Lord didn't want us to take as much as we could get... He wouldn't have given us two good hands to take it with'.[1] However, those who know about Union agricultural law know that the EU's support programmes are not a free self-service buffet for European farmers. The very complex rules show that it is a difficult balancing act to determine the strictness of the requirements that legislators and administrators can impose on the industry which receives a major portion of the European Union's budget as aid.

---

1. Joseph Heller, *Catch-22*, reprint 1967, p. 83

This book has been written to meet the needs both of university-level teaching and of those who are professionally engaged in dealing with issues of Union agricultural law, whether lawyers, professional interest groups or administrative authorities. The writing of this book has also been stimulated by surprise at the relatively little interest that appears to have been shown in agricultural law in the legal literature. As can be seen from the references to the literature in the footnotes to the text and in the bibliography, reputable books on this subject have been written by other authors, but compared with the wealth of literature on other Union law topics, agricultural law has received very little attention and this is a shame.

As a discipline, Union agricultural law takes the reader into many areas of the legal landscape. Considerations of space have meant that this book is restricted to the core areas of the agricultural rules. Topics which might well have been included are omitted or only touched on in passing. This is the case, for example, with the environmental regulation of agriculture and the rules on food safety. In some areas, Union agricultural law goes into a level of detail which it is not appropriate to describe and examine in a book such as this. Even in those areas that are dealt with, it has been necessary to be selective.

The basis for this book is my book *EU-landbrugsretten*, which was published in Danish in 2009 by Jurist- og Økonomforbundets Forlag. This book is an updated and expanded version of that book; it has been translated by Steven Harris. The publication of this translation has been made possible by the help of The Dreyer Foundation, Copenhagen, Denmark (Dreyers Fond), and I am grateful both to Dreyers Fond and to Steven Harris.

Comments and suggestions about the book and about EU agricultural law in general will be gratefully received at the following addresses:

jd@jura.au.dk
Aarhus Universitet
Juridisk Institut
Bartholins Allé 16
DK-8000 Århus C
Denmark
Aarhus, August 2012
Jens Hartig Danielsen
Professor
dr jur., PhD, cand.jur., LLM, BA

List of Abbreviations

ABl	Amtsblatt
AUR	Agrar- und Umweltrecht, Münster
BISD	Basic Instruments and Selected Documents
Bull.	Bulletin
CAP	Common Agricultural Policy
Case W. Res. J. Int'l L.	Case Western Reserve Journal of International Law, Cleveland
CDE	Les Cahiers du droit européen, Bruxelles
COM	European Commission Documents
Coreper	Permanent Representatives of the Governments of the Member States
CSA	Special Committee on Agriculture
Drake J. Agric.L.	Drake Journal of Agricultural Law
EAFRD	European Agricultural Fund for Rural Development
EAGGF	European Agricultural Guidance and Guarantee Fund
EC	European Community
EEC	European Economic Community
ECR	European Court Reports
ELR	European Law Review, London
FCLI	Fordham Corporate Law Institute
GATT	General Agreement on Tariffs and Trade

HILJ	Harvard International Law Journal
Ibid.	Ibidem
IBRD	International Bank for Reconstruction and Development
IMF	International Monetary Fund
ITO	International Trade Organization
JIEL	Journal of International Economic Law
JO	Journal officiel
NILQ	Northern Ireland Legal Quarterly
No.	Number
OJ	Official Journal of the European Union
p.	page
Rec.	Recueil de la Jurisprudence de la Cour
RIW	Recht der Internationale Wirtschaft, Heidelberg
Rn.	Randnummer
RTDE	Revue trimestrielle de droit européen
TEC	Treaty establishing the European Community
TEEC	Treaty establishing the European Economic Community
TFEU	Treaty on the Functioning of the European Union
UNTS	United Nation Treaty Series
VO	Verordnung
Vol.	Volume
WCO	World Customs Organization
WTO	World Trade Organization
YEL	Yearbook of European Law, Oxford

CHAPTER 1

The Basis and Development of EU Agricultural Law

§1.01 THE HISTORY

The great majority and the most important of the rules that fall under the heading of European Union (EU) agricultural law are found either in the Treaty on the functioning of the European Union (TFEU) or in secondary legislation adopted on the legal basis of the TFEU. The TFEU also contains the procedural principles for the institutions than can adopt EU legislation and issue administrative decisions. On the other hand, the rules for the institutional framework, i.e., the rules of the institutions themselves, are not contained in the TFEU but in the Treaty on European Union (TEU).[1]

The organization that has developed to become the EU began in 1950 with the proposal to set up the European Coal and Steel Community (ECSC). The ECSC Treaty established a European community with institutions that were independent of its Member States, with independent regulatory powers over – as was implied by its title – the areas of coal and steel production.

Among other things, World War II and the subsequent economic hardships in Europe led to major shortages of food, and in 1957 the experiences gained from the ECSC with its supranational institutions were applied in the Treaty of Rome, establishing the European Economic Community (EEC). This established a more general community and one of the most important areas of its competencies and policies concerned agriculture. In 1992 the EEC Treaty was replaced by the Treaty establishing the European Community (EC), but its provisions on agriculture remained unchanged

---

1. See §1.03 below on the significance of the Lisbon Treaty. On the concept of EU agricultural law, Leidwein, *Einführung in das europäische Agrarrecht* 2 (Berliner Wissenschaftsverlag, Österreichischer Agrarverlag and Neuer wissenschaftlicher Verlag, 1996) states that: '*[d]em Europäischen Agrarrecht sind jene Normen zuzuzählen, die funktionell gesehen hinsichtlich der Land- und Forstwirtschaft spezifische Wirkungen entfalten, mögen sie nun einem von typischen agrarischen Interessen geprägten Rechtsbereich entstammen oder einem solchen, der vorwiegend von anderen als agrarisch bestimmten Verwaltungszwecken beherrscht ist'.*

1

from the 1957 Treaty.[2] Under Union law, the legal regulation of agriculture falls under the heading of the 'common agricultural policy'(CAP).[3]

The common agricultural policy is built on three pillars: the first pillar involves farm support and market organizations; the second pillar consists of rural development policy; and the third pillar consists of legislation providing the framework for agricultural production in general, including legislation on animal husbandry and agricultural products.[4]

As with many other policy areas, the history of the common agricultural policy contains many different elements that flow into and through each other. However, it is possible to point to some milestones where there have been significant initiatives or a reordering of priorities in agricultural policy.

The basis for the common agricultural policy was laid in 1958–1960 by the six original Member States at the Stresa Conference and with the first Mansholt Plan, named after the then Commissioner for Agriculture.[5] This established that the basic instrument of the common agricultural policy was to be the common organization of agricultural markets for a number of different agricultural products.[6] The measures for the individual market sectors were – and still are – all based in varying degrees on three main principles. The first is the principle of unified markets or common prices; the price for each agricultural product is the same throughout the EU. The second is the principle of financial solidarity; the costs associated with the market organizations are borne by the EU. Finally, there is the principle of Union preference, whereby in balancing the interests of a third country and a Member State, the interests of the Member State or of the Union are given preference.[7] It is also possible to point to other principles that can either be regarded as special applications of one of the three main principles or as independent principles that are not as important as the main principles. These include the intervention principle, under which agricultural producers are guaranteed a minimum price for their products; the import protection principle, according to which agricultural production in the Union is to some extent protected

---

2. In 1997, the Treaty of Amsterdam deleted Articles 44, 45 and 47 from the part of the EC Treaty that concerned agriculture, as they had become redundant.

3. In French the 'politique agricole commune' (PAC) and in German the 'gemeinsame agrarpolitik' (GAP).

4. For a legal historic review of pillars one and two of the CAP, see Cardwell, The European Model of Agriculture 175–234 (Oxford University Press, 2004).

5. See the conclusions of the Stresa Conference of 1958 in ABl 1958 L 281/JO 1958 L 291, and the First Mansholt Plan, COM(60) 105 and Bull. CE 1/61, s. 83.

6. For further details on the development of EU agricultural law, see Barents, The Agricultural Law of the EC 3–9 (Kluwer, 1994), with a number of references; Wulff, EU-jordbrugsret, 2001, pp. 11-12; Cardwell, The European Model of Agriculture 7–174 (Oxford University Press, 2004), with a comprehensive review of the dogmatic history of EU agricultural law; Leidwein, Europäisches Agrarrecht 76-78 (Berliner Wissenschaftsverlag, Österreichischer Agrarverlag and Neuer wissenschaftlicher Verlag, Ed. 2, 2004); and McMahon, EU Agricultural Law 37–72 (Oxford University Press, 2007).

7. The three main principles of the common agricultural policy are set out in: Usher, EC Agricultural Law 39–40 (2d ed., 2001); Leidwein, Europäisches Agrarrecht 80-82 ((Berliner Wissenschaftsverlag, Österreichischer Agrarverlag and Neuer wissenschaftlicher Verlag, Ed. 2, 2004); and McMahon, EU Agricultural Law 39 (Oxford University Press, 2007).

from imports from third countries; and the export principle, under which the union supports exports of agricultural products to third countries.

Market organizations were already established for the most important agricultural products by the mid 1960s, thus establishing the first pillar of the common agricultural policy.[8] The politically determined price level ensured high prices for agricultural producers, and the means for achieving this were, in particular, intervention purchases, taxes on imported products from third countries, and refunds on exports to third countries.

In 1962, the European Agricultural Guidance and Guarantee Fund (EAGGF) was set up, with responsibility for the financing of the common agricultural policy up to 2007. The EAGGF was divided into two sections. The Guidance Section financed part of the common agricultural policy's second pillar and was similar to one of the structural funds which the Union uses, not just in the area of agricultural but in a number of areas, for promoting regional development and reducing the differences between areas within the EU. The Union provided support for major individual projects through the Guidance Section, both for production and marketing, and it also financed part of the LEADER initiative (which plays an important role in rural development by encouraging local strategies based on partnership and experience-sharing networks).[9] The Guarantee Section was much more important as it was this which financed the individual market organizations, for example for intervention buying and payment of export refunds, i.e., the first pillar of the common agricultural policy. The Guarantee Section also financed parts of the costs for the second pillar of the common agricultural policy that were not covered by the Guidance Section.[10] Since 2007, the financing of the common agricultural policy has been reorganized and it is now financed through two agricultural funds; see below.

In 1968, the Commission put forward the Second Mansholt Plan.[11] The Plan was in part the basis for the reform of the common agricultural policy which was carried out in 1972 in conjunction with the extension of the membership of the European Community. It was seen that the pricing policy had not ensured that farmers had a sufficiently high living standard, and that it led to over-production. The 1972 reform involved a strengthening of the second pillar of the common agricultural policy, introducing a structural policy giving various forms of support to agriculture. To begin with the structural policy consisted of investment aid for the modernization of farming, aid for the discontinuation of cultivation or for making land available for modernized

---

8. The market organizations for some agricultural products were first introduced much later, for example bananas in 2003 and biofuels in 2003. There have never been independent market organizations for potatoes, honey, vinegar or horses. On the current Single CMO Regulation, see Ch. 4.
9. On the LEADER initiative, which is carried forward by the Rural Development Regulation (Council Regulation (EC) No 1698/2005) see Ch. 5, §5.02[A][4].
10. Until 2007, the rules on the financing of the common agricultural policy were set out in Council Regulation (EC) No 1258/1999 on the financing of the common agricultural policy, but with effect from 2007 this has been replaced by Council Regulation (EC) No 1290/2005.
11. See Le Roy, *La Politique Agricole Commune 35-38* (Economica 1994), on the background to the plan.

use, and aid for advice.[12] In the same period, a quota arrangement was introduced for milk production. When structural aid was subsequently found not to have the desired effect for the position of smaller agricultural holdings, and since it led to intensive farming that was harmful to the environment, the balance of structural aid was later changed. Aid for investment was now targeted at environmental improvements and improvements in energy use as well as for taking agricultural land out of production.[13]

In 1992, a reform was implemented which was later called the MacSharry Reform, after the then Commissioner for Agriculture. The reform was the hitherto biggest change to the common agricultural policy. The main aim of the reform was to lower agricultural prices in order to make the Union's agricultural products more competitive, both in the internal market and on the world market, at the same time as compensating agricultural producers for loss of income. The reform reduced the importance of the price supports for certain agricultural products that had been paid hitherto, and instead gave support directly to the producers. A number of the prices that had been set by politicians were lowered, and support was paid on a per-hectare basis or a per animal basis. As a special production restriction mechanism, the reform introduced a requirement for larger producers to set aside a certain proportion of their land.[14]

At the end of the 1990s, the Agenda 2000 strategy was drawn up for the EU's future policies in connection with the expansion of the Union with the admittance of a number of Central and Eastern European countries.[15] In the area of agriculture, the strategy built further on the MacSharry Reform, involving a thoroughgoing reform of the common agricultural policy. As something new, the Agenda 2000 strategy put a cap on expenditures for the common agricultural policy.[16] The importance of price supports was further reduced for a number of market organizations; for example the guaranteed minimum price under the market organization for beef was reduced by 20%, and there was a reduction of 15% for arable products, milk and dairy products.

The second important element was the implementation of a thorough reform of the structural policy. The structural policy that had been pursued hitherto was replaced by a rural districts policy, though to a large extent it carried forward the structural support and accompanying measures of the MacSharry Reform of 1992. The reform meant that the second pillar of the common agricultural policy was now called the rural development policy. The provisions on rural development policy are in the Rural

---

12. Cardwell, *The European Model of Agriculture*, 26–28 (Oxford University Press, 2004).
13. See the Commission's proposal for a fundamental reform in which there should be an attempt to match supply and demand by introducing new methods for restricting production in certain sectors: Communication from the Commission to the Council and the Parliament – Perspectives for the common agricultural policy (Green Paper), COM(85) 333 final.
14. For criticisms of the MacSharry Reform see: Schulze, *Ratgeber Agrarreform 16-18* (Landwirtschaftsverlag, Ed. 2, 1998).
15. On the special agricultural rules for the Eastern European Member States see: Cardwell, *The European Model of Agriculture* 388–404 (Oxford University Press, 2004).
16. Already in 1988, the European Council had laid down agricultural guidelines to limit the percentage of total Union expenditure that could be devoted to the agricultural policy.

Development Regulation.[17] The Regulation set out a seven-year programme for 2000–2006, aimed at coordinating the interests of agriculture with the interests of the rural economy in general. The aim was to coordinate overall commercial considerations with considerations for the environment, nature and landscape. Apart from the part of the rural development policy that concerned environmentally friendly agricultural measures, the Member States had considerable freedom to choose which support measures they would implement.

Even though, as stated, the intention was that the Agenda 2000 strategy would apply up to 2006, the Commission nevertheless decided to put forward a proposal for a reform of the common agricultural policy that was later referred to as the 2003 Reform.[18] The most important element in this proposal was the intention to promote a more market-oriented and sustainable agricultural sector by continuing to convert agricultural support from product support (price support) to direct support for producers. This was to be achieved by means of decoupled support; this meant that agricultural support was to be paid directly to the individual farmer without regard for the quantity of goods produced. The Commission emphasized that if the competitiveness of the Union's agriculture was to be improved at the same time as promoting more sustainable farming methods (better food quality and less damage to the environment), this would require both a reduction of the institutional prices (price support through market organizations) and an increase in production costs for agricultural units in the Union. As for the financial costs for the Union for changing from production support to decoupled support, the Commission stated that the decoupling would not lead to any alteration of the amount in fact paid to support agriculture in the Union.[19]

The decoupling of agricultural support from production has not taken place all at once or at the same time for all market organizations. It is planned that decoupling should be applied to the whole of the arable sector, while aid for animal husbandry will still be linked in part to production.

The 2003 Reform also introduced a new instrument of the common agricultural policy in the form of 'dynamic modulation'. This refers to arrangements where the direct payments to farms are gradually reduced by a certain percentage each year until the targeted reduction is reached. Within the common agricultural policy the development of rural districts is strengthened by the transfer of money from the first pillar (market organizations) to the second pillar (rural development policy) by a system of compulsory dynamic modulation and an expansion of the possibilities for rural

---

17. Council Regulation (EC) No 1698/2005 on support for rural development by the European Agricultural Fund for Rural Development (EAFRD), (OJ L 277, 21 Oct. 2005, 1).
18. Communication from the Commission to the Council and the European Parliament – – Mid-Term Review of the Common Agricultural Policy, COM(2002) 394 final. On the 2003 Reform see: Ahner in AUR 2005, No 8, Annex: *Europäisches Agrarrecht*, p. 3.
19. See COM(2002) 394 final, recital 22. Recital 22 is repeated as recital 24 in Council Regulation (EC) No 1782/2003 establishing common rules for direct support schemes under the common agricultural policy and establishing certain support schemes for farmers. See also Peter Mortensen, *Landbrugets retsforhold II, Landbrugsloven*, 59-62 (Pejus, 2005); and on the 2003 Reform and its implementation in Denmark, see Jens Evald, *Landbrugets retsforhold III – Landbrugsrelateret ret* 68(Jurist- og Økonomforbundets Forlag 2007).

development with a view to improving food quality, fulfilling higher norms and promoting animal welfare.[20]

As stated above, with effect from 2007 the previous Sections for the financing of the common agricultural policy (the Guidance Section and the Guarantee Section) have been terminated and instead two agricultural funds have been set up, each financing their pillar of the common agricultural policy. The European Agricultural Guarantee Fund (EAGF) finances the first pillar of the common agricultural policy, primarily the market organizations.[21] The other fund, the European Agricultural Fund for Rural Development (EAFRD), finances the rural development policy, i.e., second pillar of the common agricultural policy.[22]

In 2008, the EU was among the world's biggest importer and exporter of agricultural products. The international regulation of trade in agricultural products thus has a significant influence on the development of the common agricultural policy. The GATT agreements and the establishment of the World Trade Organization (WTO) in 1995 made its members, including both the EU as such and the EU Member States, subject to certain requirements in framing their local agricultural policies. The WTO rules have meant that the EU has had to convert variable import duties into fixed duties, to reduce export subsidies, both in respect of the volume of exports and the amount of the subsidies, and to reduce the general level of subsidies. The WTO rules were a significant reason for the introduction of the 2003 Reform of the common agricultural policy.

Since the mid 1990s, the third pillar of the common agricultural policy (legislation on the framework for agricultural production in general) has played an ever more important role. In 2000, the Commission published a White Paper on food safety,[23] and developments continue for Union regulation of topics such as food safety, food quality, animal welfare and environmental harm.

§1.02   THE POSITION OF THE COMMON AGRICULTURAL POLICY

The TFEU is divided into seven parts, each of which is divided in turn into a number of titles. Part Three is headed 'Union policies and internal actions', and its Title III, which covers Articles 38–44, is headed 'Agriculture and fisheries'. These provisions are often referred to collectively as the 'common agricultural policy'.

Article 38(1) states that the internal market extends to agriculture, fisheries and trade in agricultural products. However, Article 38(2) makes it clear that it is possible to make special exceptions for agriculture; it states that the rules laid down for the establishment and functioning of the internal market apply to agricultural products, unless otherwise provided for in Articles 39–44.

Article 39 sets out the objectives of the common agricultural policy, while Articles 40 and 41 define the forms of regulations and other measures that can be used to

---

20.   COM(2003) 23 final, p. 3.
21.   Article 3 of Council Regulation (EC) No 1290/2005 on the financing of the common agricultural policy.
22.   *Ibid.*, Art. 4.
23.   White Paper on food safety, COM(1999) 719 final.

achieve these objectives.[24] Article 43 gives the European Parliament and the Council the power to adopt legal acts in the form of regulations, directives, decisions and recommendations. As in nearly all other policy areas, the Commission has a monopoly on taking initiatives, as only the Commission can make proposals for regulations etc.[25] Article 42 establishes that the provisions of the Chapter relating to rules on competition, which include the provisions on State aid, only apply to production of and trade in agricultural products to the extent determined by the European Parliament and the Council. The final provision in the Title on agricultural policy, Article 44, only applies to the few situations where a product in a Member State is subject to a national market organization. In practice, Article 44 only applies to coffee, cork and horsemeat.[26]

The common agricultural policy is on the same footing as other policy areas with regard to the rules on the internal market, including the free movement of goods, workers, services and capital, as well as the rules on competition. However, the common agricultural policy is without question the most economically significant policy area in Union law, as well as the area in which Union regulation has been implemented most consistently and intensely.

§1.03 THE LISBON TREATY AND THE COMMON AGRICULTURAL POLICY

The Lisbon Treaty changed the heading of the TFEU's rules on agriculture from 'Agriculture' to 'Agriculture and fisheries', but did not directly make great substantive changes.

One important change is that under the 'ordinary legislative procedure' referred to in Article 294 TFEU, which corresponds to the co-decision procedure in Article 251 of the Treaty establishing the European Community (EC Treaty), Article 43(2) TFEU now applies in connection with the adoption of rules on market organizations and on the realization of the goals of the common agricultural policy. As for the fixing of prices, levies, aid and quantitative limitations and on the fixing and allocation of fishing opportunities, the legislative procedure laid down in the EC Treaty has been retained, so that the European Parliament must merely be consulted by the Council.

As something new, Article 13 TFEU emphasizes that the formulation and implementation of the Union's policies on agriculture, fisheries and transport should pay full regard to the welfare requirements of animals, 'since animals are sentient beings'.

---

24. See further in Ch. 2, §2.05.
25. On the legislative procedure for the common agricultural policy, see Ch. 3, §3.05.
26. See, among others, Leidwein, *Europäisches Agrarrecht 76 ((Berliner Wissenschaftsverlag, Österreichischer Agrarverlag and Neuer wissenschaftlicher Verlag,*, Ed. 2, 2004).

CHAPTER 2
Basic Concepts, Aims and Considerations

§2.01 'AGRICULTURE'

The common agricultural policy is concerned with 'agriculture', but Union law does not contain a definition or delimitation of what is covered by the term.[1] However, the provisions on the common agricultural policy in the TFEU, and Union legislation adopted pursuant to the TFEU, show that the term 'agriculture' includes what can be considered primary agriculture,[2] processing and distribution undertakings,[3] as well as agricultural training and research.[4] In the context of Union law, fisheries also fall under the heading of agriculture, but forestry does not.[5] The fact that the TFEU does not contain a more specific definition of the scope of the common agricultural policy means that it is largely left to EU legislator to define the scope of the policy area. Union legislation thus becomes an interpretation of the TFEU.[6]

In Council Regulation (EC) No 73/2009 on direct support schemes for farmers etc., 'agricultural activity' is defined as the production, rearing or growing of agricultural products including harvesting, milking, breeding animals and keeping animals for farming purposes, or maintaining the land in good agricultural and environmental condition.[7]

---

1. Case 85/77 *Sant'Anna* [1978] ECR 527, para. 8; and Case 139/77 *Denkavit* [1978] ECR 1317, para. 11.
2. See e.g., Art. 39(1)(a) TFEU.
3. See e.g., Art. 38(2) TFEU.
4. See Art. 41(a) TFEU; and Cardwell, *The European Model of Agriculture*, 235–238 (Oxford University Press, 2004).
5. On fisheries, see Art. 38(1) TFEU. 'Timber' is not referred to in Annex I to the TFEU and is thus not an agricultural product.
6. See Jens Hartig Danielsen, *Parallelhandel og varernes frie bevægelighed* 22 (Jurist- og Økonomforbundets Forlag, 2005) on the EU legislator' role as interpreter of the Treaties.
7. Council Regulation (EC) No 73/2009 establishing common rules for direct support schemes for farmers under the common agricultural policy and establishing certain support schemes for farmers, etc. (OJ L 30, 31 Jan. 2009, 16)

§2.02   'AGRICULTURAL PRODUCTS'

There are several reasons why it is worth defining 'agricultural products'. Agricultural products, as a category, have a special status in Union law. First, not all the ordinary provisions of the TFEU apply to them. Second, there is a long list of special rules that only apply to agricultural products. Finally, the categorization of a good is relevant to where the legal basis for regulating certain matters concerning the good is to be found within the TFEU.

As mentioned in the preceding section, the TFEU only gives an outline of what 'agriculture' means, but there is more substance to the Treaty definition of agricultural products. In Article 38(1) TFEU, it is stated that: 'Agricultural products' means the products of the soil, of stockfarming and of fisheries and products of first-stage processing directly related to these products.' This definition is exemplified by the reference in Article 38(3), where it is stated that the products subject to the provisions of Articles 39–44 are listed in Annex I to the TFEU. The list in Annex I is based on the Brussels nomenclature, and it gives an exhaustive list of agricultural products.[8] If a good cannot be interpreted as being listed in the Annex, it can only be treated as an agricultural product within the meaning of the Treaty if there is an amendment to Annex I by means of an amendment to the Treaty.[9]

However, if one compares the definition of agricultural products in Article 38(1) with the exhaustive list in Annex I, it quickly becomes clear that they are not in total agreement.[10] For example, the definition in Article 38(1) does not include cheese or butter, or margarine or other edible fats, because these are products that have been subject to more than 'first-stage processing'. However, cheese and butter are covered by the term 'dairy produce' in Annex I, and Annex I expressly refers to margarine and 'other prepared edible fats'. However, the list in Annex I does not include agricultural

---

8.   The nomenclature used in Annex I to the TFEU is based on the Convention on Nomenclature for the classification of goods in Customs tariffs, which entered into force in 1959, which in turn was based on the work of the former Customs Cooperation Council. In 1994 this organization changed its name to the World Customs Organization (WCO). The European Union's common customs tariff is based on the nomenclature of the newer Convention on the Harmonized Commodity Description and Coding System of 1988. There does not appear to be any good reason why these different nomenclatures should be used in Union law, and the confusion is only increased by the fact that more recent legislation governing agriculture has clearly used the 1988 nomenclature. Use of the 1988 nomenclature at Treaty level within the area of the common agricultural policy would require an amendment to the Treaty (amendment of its Annex I). The proposed Treaty establishing a Constitution for Europe used the 1959 nomenclature, as hitherto, and the Lisbon Treaty did not alter this.
9.   See the contrary view in Constantinides-Megret, *La politique agricole commune en question* 5 (A. Pedone, 1982) who argued that the list in Annex I could be amended on the legal basis of Art. 308 of the EC Treaty (now Art. 352 TFEU). In support of the argument made here, that an amendment to the list in Annex I would require a Treaty amendment, see Barents, *The Agricultural Law of the EC* 27 (1994).
10.   See Barents, *The Agricultural Law of the EC* 27 (1994), for an explanation of the disagreement between the definition in Art. 38(2) and the list which refers to Art. 38(3).

products that are clearly covered by the definition in Article 38(1), such as timber, skins, cotton and wool.[11]

In Case 185/73 *König*, the Court of Justice interpreted the definition of agricultural products in Article 38(1) and (3). Originally, the Council had had power to add products to the list in Annex I for a period of two years following the entry into force of the EEC Treaty.[12] The Council had used this power to add ethyl alcohol to the list, among other things, and the case concerned whether this addition was or was not permitted, as it was argued that ethyl alcohol was not an agricultural product.[13]

The Court of Justice pointed out that the definition of agricultural products is placed at the head of the title on agriculture. It found that this definition would be devoid of practical meaning if, as regards the power of the Council to fill the gaps during the two-year period referred to, it were not to be interpreted 'in the light of the aims of the common agricultural policy and with reference to the products with which the authors of the treaty considered that policy to be concerned'.[14] While the Court emphasized that this interpretation related to the Council's special two-year power, on the basis of the judgment it must be assumed that the definition of agricultural products must generally be based on a combination of Article 38(1) and (3). However, even prior to the *König* judgment it had been established that in the event of a conflict between the definition of agricultural products and the list, then the list should take priority.[15] This means that if a product is named in the Annex I list, but does not fall within the definition in Article 38(1), then it is considered to be an agricultural product for the purposes of the TFEU. The priority given to the list also means that if a product is covered by the definition but is not listed, then it is not considered to be an agricultural product. In other words, the interpretation of the Annex I list is decisive for whether a product is to be treated as an agricultural product or not. The function of the definition in Article 38(1) is to guide the interpretation of the list in Annex I.[16]

Another important aid to the interpretation of the Annex I list is the explanatory notes on the International Convention on the Harmonized Commodity Description and Coding System which forms the basis for Annex I.[17] In Case 61/80 *Coöperatieve Stremsel- en Kleurselfabriek*, the Court of Justice stated, a little categorically, there are no Union provisions explaining the concepts contained in Annex I, and that as Annex I adopts word for word certain headings of the Customs Cooperation Council nomenclature, 'it is appropriate to refer to the explanatory notes on the nomenclature in order

---

11. In the case of cotton, a special arrangement was made upon the entry of Greece into the European Communities; see the Act concerning the conditions of accession of the Hellenic Republic and the adjustments to the Treaties (OJ L 291, 19 Nov. 1979,174).
12. Art. 38(3) of the EEC Treaty.
13. Ethyl alcohol and other products were included in the list in Annex I by Council Regulation 7a/59 of 18 December 1959 (OJ 1961, 71).
14. Case 185/73 *König* [1974] ECR 607, para. 13.
15. In Joined Cases 2 and 3/62 *Commission v. Luxembourg and Belgium*, ECR English special edition, p. 445, the Court had ruled that the list in Annex I should be considered exhaustive.
16. Case 77/83 *CILFIT* [1984] ECR 1257, paras 11–12. Snyder, *Law of the Common Agricultural Policy* 17 (Sweet & Maxwell, 1985), stated that in practice Art. 38(1) has been interpreted as being 'merely indicative'.
17. See footnote 8 above.

to interpret the Annex'.[18] In the *CILFIT* case, the Court went further and stated that since there are no Union provisions explaining the concepts contained in Annex I and as the Annex adopts word for word certain headings of the Common Customs Tariff, the established interpretations and methods of interpretation for the Common Customs Tariff should be used to interpret the Annex.[19]

However, with regard to Article 38(1) and (3), it is not only the definition in Article 38(1) that influences the interpretation of the list in Annex I. The list has also influenced the interpretation of the reference in Article 38(1) to 'products of first-stage processing directly related to these products'.[20] In the *König* case, the Court pointed out that the list in Annex I not only includes the principal agricultural products, but also a certain number of foodstuffs whose remoteness, in industrial terms, from the basic agricultural product goes beyond the point of first-stage processing as understood in a restricted sense. The Court ruled that the number of operations necessary to obtain a processed product is not the criterion for determining whether a product is covered by Article 38(1). According to the Court, the element that is common to these products resides in the close economic interdependence between them and the basic products, so that it would not be justifiable to apply the rules of the common agricultural policy to the basic products while applying the general rules of the Treaty to the processed products. The Court ruled that the wording 'products of first-stage processing directly related to these products' means that there must be a clear economic interdependence between basic products and products resulting from a productive process, irrespective of the number of operations involved therein. The Court also stated that processed products which have undergone a productive process, the cost of which is such that the price of the basic agricultural raw materials becomes a completely marginal cost, are not covered by the rules of agricultural law.[21]

As stated above, the *König* case concerned ethyl alcohol which was produced on the basis of sugar, an agricultural product. The Court of Justice found that ethyl alcohol is a product which, for the purposes of Article 38(1), is directly related to an agricultural product and which has undergone first-stage processing.

This ruling was neither altered by the fact that ethyl alcohol with a strength of less than 80% by volume would, in practice, be subject to an additional process after distillation of dilution with water (which is not an agricultural product).[22]

In the *Coöperatieve Stremsel* case, the Court of Justice ruled that animal rennet is not an agricultural product. Rennet is not listed in Annex I and could not be regarded as being covered by the list even though rennet is used in the production of products (cheeses) which are on the list.[23] The case concerned the interpretation of a regulation

---

18. Case 61/80 *Coöperatieve Stremsel- en Kleurselfabriek*, [1981] ECR 851, para. 20.
19. Case 77/83 *CILFIT* [1984] ECR 1257, para. 7. On the relationship between the nomenclature and Union law, see Usher, Com. Mkt. L. Rev, 389 (1982).
20. Barents, *The Agricultural Law of the EC* 27, n. 92 (1994), does not ascribe much practical significance to the effects of Art. 38(1) and (3), stating that Annex I 'remains decisive'.
21. Case 185/73 *König* [1974] ECR 607, paras. 12 and 14.
22. *Ibid.*, para. 10.
23. Case 61/80 *Coöperatieve Stremsel- en Kleurselfabriek*, [1981] ECR 851, para. 21. See also Case C-250/92 *DLG* [1994] ECR I-5641, para. 23.

which expressly referred to the list in Annex I, and the Court therefore did not have to determine whether the definition of a processed product in Article 38(1) can in itself mean that a product is an agricultural product.

To the extent that an agricultural product is covered by the definition in Article 38(1), but is not listed in Annex I to the Treaty, in principle the legal basis for its regulation under Union law is not to be found in Article 43 but elsewhere in the Treaty, for example Article 352.[24] However, this starting point is departed from in several cases. With regard to the definition of the competence of the EU legislator pursuant to Article 43 TFEU, it is essential that the legislation should be necessary for the pursuit of the objectives of the common agricultural policy which are set out in Article 39, and that the legislation should be directly or indirectly linked to one or more agricultural products within the meaning of Annex I. The requirement that the legislation must be directly or indirectly linked to an agricultural product means that the legislation must concern the production or sale of agricultural products and thereby fulfil the objectives in Article 39 TFEU.[25] → the competence link can be indirectly linked to agricultural product

In the *Dansk Landbrugs Grovvareselskab* case, the Court of Justice accepted that Article 43 can constitute the legal basis for legislation that does not directly concern agricultural products, in this case Council Directive 91/414/EEC on the placing of plant protection products on the market.[26] Plant protection products are not agricultural products, see Article 38(3) TFEU, but the use of plant protection products is one of the most important means for protecting plants and plant products and thus for improving agricultural production.[27] Increased agricultural productivity is one of the objectives of the common agricultural policy.[28]

§2.03   'AGRICULTURAL PRODUCER'

The common agricultural policy contains a great number of rules and regulations that are directed at agricultural producers. In particular, in the area of rural development policy there are rules and regulations on for example investment support, advice and pensions. However, the TFEU does not contain any definition of an 'agricultural

---

24. In footnote 11 above there is a reference to arrangements for cotton which was originally regulated under the Act concerning the conditions of accession of the Hellenic Republic and the adjustments to the Treaties (OJ L 291, 19 Nov. 1979, 174), but which is now governed by Council Regulation (EC) No 73/2009. See Ch. 5, §5.02[B], on the use of Art. 43 TFEU as the legal basis for the adoption of strategic guidelines for rural development policy.
25. Barents, *The Agricultural Law of the EC*, 1994, 179. However, see Barents 29, n. 101; and McMahon, *EU Agricultural Law* 6 (Oxford University Press, 2007), which both refer to Case 123/83 *BNIC* [1985] ECR 391 as a precedent for the view that Art. 43 TFEU cannot be the legal basis for acts which directly concern products not listed in Annex I. Apparently, however, the only thing the Court of Justice established in this case was that potable spirits are expressly excluded from the category of agricultural products and are thus industrial products, and this is not altered by the economic importance of the products for farmers in the region concerned (Cognac); see paras 14 and 15.
26. Case C-250/92 *DLG* [1994] ECR I-5641, para. 26.
27. Regulation (EC) No 1107/2009 of the European Parliament and of the Council of 21 October 2009 concerning the placing of plant protection products on the market (OJ L 309, 24 Nov. 2009, 1), recital 6.
28. Article 39(1)(a) TFEU.

producer'.[29] As a consequence it is largely up to the EU legislator to determine the scope of the concept, and as this is typically in the form of legislation on specific situations, the scope of the concept can vary according to the context.[30] In the *Denkavit* case the plaintiff in the main case argued that the concept of an agricultural producer includes all producers of agricultural products within the meaning of Article 38 TFEU and Annex I, but the Court rejected this definition. The Court stated that the concept of agriculture is not precisely defined in the Treaty and that it is thus 'for the competent authorities where necessary to define the scope of such rules in relation to persons and in relation to subject-matter'.[31] In Union legislation it can be left to the Member States to determine the concept of an agricultural producer under national law.

The nature of the discretion which sometimes is left to the Member States is illustrated by two cases. The first case, the *Denkavit* case, concerned a Regulation that authorized Germany to give aid 'in the form of direct aids to agricultural producers' with a view to compensating for losses caused by the reduction in prices for agricultural products. The German legislator used this authorization among other things to adopt a law giving support to agricultural activities involving animal husbandry and cattle stockholding, where the support was linked to the area of agricultural land used. Among other things the Denkavit company's activities included the fattening of calves by means of substitute milk-based fodder, and it did not use agricultural land linked to the fattening of the calves. Denkavit was thus not entitled to support pursuant to the German legislation. The Court of Justice gave the concept of an agricultural producer a specific interpretation and found that:

> since neither the context nor the objectives of the regulation demand a restrictive interpretation, it is not out of the question that the relatively broad expression 'agricultural producers', which is used in the wording of the regulation, may include production of agricultural products by any method whatever.

It did not appear that the Regulation imposed any obligation, but merely authorized the German authorities to give support. Therefore, the Regulation did not prohibit Germany from excluding industrial breeders/stockholders from receiving the support referred to in the Regulation.[32]

The second case, the *Monte Arcosu* case, concerned the Union law rules concerning persons who have farming as their main occupation.[33] According to the Regulation on improving the efficiency of agricultural structures then in force, aid could be paid for investment in agricultural holdings where the farmer practised

---

29. Barents, *The Agricultural Law of the EC* 26 (1994), states that the reason for the absence of a Treaty definition of, among other things, an 'agricultural producer' was that: 'the common agricultural policy was primarily conceived as a means to extend the free movement of goods to the agricultural sector'.
30. Barents, *The Agricultural Law of the EC* 31 (1994); and McMahon, *EU Agricultural Law* 3 (Oxford University Press, 2007).
31. Case 139/77 *Denkavit Futtermittel* [1978] ECR 1317, paras 10 and 11.
32. *Ibid.*, paras 12–13.
33. Case C-403/98 *Monte Arcosu* [2001] ECR I-103. On farming as a part-time occupation, see Case 152/79 *Lee* [1980] ECR 1495; and Case 107/80 *Adorno* [1981] ECR 1469.

farming as his main occupation.[34] Within certain limits the Regulation left it to the Member States to define what is meant by the expression 'farmer practising farming as his main occupation'. In the case of a natural person, the proportion of income derived from the agricultural holding had to be 50% or more of the farmer's total income, and the working time devoted to work unconnected with the holding had to be less than half of the farmer's total working time. For legal persons, the Member States had greater latitude for the definition, as long as they took account of the criteria that applied to natural persons. The Court of Justice ruled that the regulation's provisions on a legal person's right to support anticipated implementing measures adopted by the Member States, and noted that such provisions had not been adopted in the Member State in question. A limited company practising farming, therefore, could not rely on the regulation before a national court.[35]

The legislation in force today provides a number of different definitions of an 'agricultural producer'. In Article 2(a) of Council Regulation (EC) No 73/2009 establishing common rules for direct support schemes for farmers etc., for the purposes of the Regulation a farmer is defined as: 'a natural or legal person, or a group of natural or legal persons, whatever legal status is granted to the group and its members by national law, whose holding is situated within Community territory... and who exercises an agricultural activity'.[36]

In connection with the rules regulating milk production, Council Regulation (EC) No 1234/2007 establishing a common organization of agricultural markets defines a 'producer' as: 'a farmer with a holding located within the geographical territory of a Member State, who produces and markets milk or who is preparing to do so in the very near future'.[37]

In some Union legislation, a distinction is made between production and processing, and it can be relevant whether a person or undertaking produces agricultural products or processes them. The Court of Justice addressed this issue directly when interpreting the relevant rules in the *Adorno* case.[38] The Commission rejected an application from a wine-grower for aid pursuant to Regulation (EEC) No 355/77 on common measures to improve the conditions under which agricultural products are processed and marketed.[39] The farmer wanted to establish a new wine-making centre on the basis of two existing holdings. The centre was intended to improve the processing into wine of the grapes produced on the farms, to rationalize the storage and preservation of the wine, to improve transport between the two farms and to shorten

---

34. Council Regulation (EEC) No 797/85 on improving the efficiency of agricultural structures (OJ L 93, 30 Mar. 1985, 1)
35. Case C-403/98 *Monte Arcosu* [2001] ECR I-103, paras 26–29.
36. Council Regulation (EC) No 73/2009 establishing common rules for direct support schemes for farmers under the common agricultural policy and establishing certain support schemes for farmers etc. (OJ L 30, 31 Jan. 2009, 16). 'Agricultural activity' is discussed in §2.01 above, and defined in Art. 2(c) of the Regulation.
37. Article 65(c) of Council Regulation (EC) No 1234/2007 establishing a common organization of agricultural markets and on specific provisions for certain agricultural products (Single CMO Regulation) (OJ L 299, 16 Nov. 2007, 1).
38. Case 107/80 *Adorno* [1981] ECR 1469.
39. OJ L 51, 23 Feb. 1977, 1.

the marketing channels for the wine, at the same time improving the quality, presentation and market preparation of the product. The Commission rejected the application on the grounds that the application fell within the scope of a Directive on the modernization of farms.[40] According to Regulation (EEC) No 355/77, aid could not be given for projects for the improvement of the processing of agricultural products if they were covered by the rules of the Directive. Under the Regulation aid could be given for the improvement of the processing of agricultural products, while the Directive concerned the improvement of the production of such products. The Court overruled the Commission's decision, since the purpose of the project submitted by the applicant was not to raise the profitability of his farm by improving production conditions for agricultural products but to improve the processing and marketing of those products, and thus it did not fall within the scope of the Directive.[41]

§2.04 'AGRICULTURAL HOLDING'

As with the term 'agricultural producer', there is no Treaty definition of the term 'agricultural holding'. In the *Sant'Anna* case, the Court of Justice found that: 'it is impossible to find in the provisions of the treaty or in the rules of secondary community law any general uniform definition of "agricultural holding" universally applicable in all the provisions laid down by law and regulation relating to agricultural production'.[42]

The definition in Article 2(b) of Council Regulation (EC) No 73/2009 establishing common rules for direct support schemes for farmers etc., is an example of a definition of an agricultural holding in Union law. The Regulation defines a 'holding' as: 'all the production units managed by a farmer situated within the territory of the same Member State'.[43] Council Regulation (EC) No 1234/2007 establishing a common organization of agricultural markets uses, by reference to it, the same definition of an 'agricultural holding' as Council Regulation (EC) No 1782/2003 establishing common rules for direct support schemes under the common agricultural policy and establishing certain support schemes for farmers.[44]

---

40. Council Directive 72/159/EEC on the modernization of farms (OJ L 96, 23 Jun. 1972, 1).
41. Case 107/80 *Adorno* [1981] ECR 1469, paras 21–23.
42. Case 85/77 *Sant'Anna* [1978] ECR 527, para. 14.
43. Article 2(b) of Council Regulation (EC) No 73/2009 establishing common rules for direct support schemes for farmers under the common agricultural policy and establishing certain support schemes for farmers etc. (OJ L 30, 31 Jan. 2009, 16).
44. Article 65(d) of Council Regulation (EC) No 1234/2007 establishing a common organization of agricultural markets and on specific provisions for certain agricultural products (Single CMO Regulation) (OJ L 299, 16 Nov. 2007, 1). Council Regulation (EC) No 1782/2003 was repealed by Art. 146 of Council Regulation (EC) No 73/2009. Pursuant to Art. 146(2), references in other legal acts to Regulation (EC) No 1782/2003 should be construed as references to Regulation (EC) No 73/2009.

§2.05 THE AIM OF THE COMMON AGRICULTURAL POLICY

[A] The Relationships between the Aims and the Discretion of the Institutions

Prior to the entry into force of the Lisbon Treaty, Part One of the Treaty establishing the European Community (EC Treaty), on the Principles of the Community, stated that one of the purposes of the Community was the introduction of a common policy in the sphere of agriculture and fisheries. By reference to Article 2 of the EC Treaty, the common agricultural policy was in itself one of the means whereby the primary goals of the Community were to be achieved. This has also been emphasized by the Court of Justice. In a case concerning the common organization of the market for wine, the Court stated that the purpose of a provision in the regulation in question was 'to achieve the objective of the rules governing the matter and consequently that of Article 39 of the EEC Treaty, which constitutes an objective of general interest pursued by the Community'.[45] This prominent position has now been amended by the Lisbon Treaty. Article 4(2) TFEU states that there is shared competence between the Union and the Member States in the area of agriculture, and Article 13 states that in formulating and implementing the Union's policies on agriculture, among other things, there must be full regard to the welfare of animals, as animals are sentient beings.

Following the Lisbon Treaty, the most important contribution to the understanding of the concept of the common agricultural policy is Article 39 TFEU. The first paragraph of the Article lists five objectives, and the second paragraph lists three considerations which must be taken into account when implementing the common agricultural policy:

Article 39 TFEU:
1. The objectives of the common agricultural policy shall be:
(a) to increase agricultural productivity by promoting technical progress and by ensuring the rational development of agricultural production and the optimum utilisation of the factors of production, in particular labour;
(b) thus to ensure a fair standard of living for the agricultural community, in particular by increasing the individual earnings of persons engaged in agriculture;
(c) to stabilise markets;
(d) to assure the availability of supplies;
(e) to ensure that supplies reach consumers at reasonable prices.
2. In working out the common agricultural policy and the special methods for its application, account shall be taken of:
(a) the particular nature of agricultural activity, which results from the social structure of agriculture and from structural and natural disparities between the various agricultural regions;
(b) the need to effect the appropriate adjustments by degrees;
(c) the fact that in the Member States agriculture constitutes a sector closely linked with the economy as a whole.

Various attempts have been made to systematize the objectives. With reference to economic theory it has been argued that the five objectives stated in paragraph 1

---

45. Case 116/82 *Commission v. Germany* [1986] ECR 2519, para. 28.

express three prime objectives that are generally used as the basis for the public regulation of agriculture. The first concerns what is called the political-economic factor, that the policy should contribute to the economic growth of the Union and its Member States. The second is the socio-political factor, whereby those involved in agriculture should be ensured a reasonable standard of living. Finally, there is the socio-economic factor, which involves ensuring a sufficient supply of foodstuffs.[46] The first prime objective, of economic growth, is expressed in Article 39(1)(a) and (c), concerning increased productivity and stabilization of the markets, and in Article 39(2)(c) referring to the central role of agriculture for the economy as a whole. The second prime objective, that of ensuring the living standards for people in the agricultural sector, is expressed particularly in Article 39(1)(b) and (2)(a) and (b), which refer to the special nature of agricultural occupations resulting from the social structure of agriculture, and the need to make the appropriate adjustments gradually. The third and last prime objective, of security of supply, is clearly expressed in Article 39(1)(d) and (e).

There have also been proposals to divide the objectives into two groups, concerning social aims and economic aims. Such a classification should support the idea that the social objectives in Article 39(1)(b) and (e), of securing a reasonable living standard for those engaged in agriculture and reasonable consumer prices for agricultural products, are the ultimate aims of the agricultural policy. In this interpretation the economic objectives referred to in Article 39(1)(a), (c) and (d) are only intermediate objectives to support the achievement of the social aims.[47]

Whether one chooses to consider the objectives in Article 39 as a number of individual main objectives or as two groups of hierarchically ranked objectives does not alter the fact that the objectives are mutually incompatible.[48]

Early decisions of the Court of Justice could be seen as supporting the idea that the objectives have a hierarchical ranking. In the *Toepfer* case, decided in 1965, a Commission decision had approved Germany's introduction of protective measures on the basis of the price development in the market for maize. Thereafter, the German authorities refused to issue import licenses for maize. One of the questions in the case was whether there was a risk of serious disturbances of the market which could threaten the achievement of the objectives set out in Article 39. The Court found that the disturbances contemplated by the Commission would have been of too temporary a nature to be capable of jeopardizing the stability of the market in maize and barley

---

46. McMahon, *Law of the Common Agricultural Policy* 24 (Pearson Education, 2000) referring to El-Agraa, *The Economics of the Common Market* 211–212 (Harvester Wheatsheaf, 4th ed., 1994); and Marsh & Swanney, *Agriculture and the European Community* 12–16 (Allen & Unwin, 1980). There is also a similar classification in Constantinides-Megret, *La politique agricole commune en question* 5-6(A. Pedone,1982).
47. Barents, *The Agricultural Law of the EC* 32 (Kluwer, 1994). Constantinides-Megret, *La politique agricole commune en question* 6 (A. Pedone, 1982) considered whether the objectives of the common agricultural policy have a special status in relation to other objectives, such as environmental protection.
48. El-Agraa, *The Economics of the Common Market* 211–212 (Harvester Wheatsheaf, ed. 4, 1994), argues that the objectives in Art. 39 interfere with the economy's natural development.

and thus of jeopardizing 'the fair standard of living for the agricultural community' mentioned in Article 39 of the [EEC] Treaty'.[49]

The special reference to the fair standard of living for the agricultural community in the *Toepfer* case could be regarded as expressing that the social objective had a special role in relation to the other objectives in Article 39. However, in a number of cases the Court of Justice has emphasized that Article 39's objectives are not hierarchical and that the Union legislator has discretionary powers to prioritize the various objectives according to the context.[50] In the *Beus* case, the Court referred to the fact that under Article 40(2), in order to establish common market organizations it is permitted to adopts provisions 'in order to attain the objectives set out in Article 39' and that these may 'include all measures required' for that purpose. On this basis the Court stated in general that the objectives of safeguarding the interests of both farmers and consumers 'may not all be simultaneously and fully attained', and that the Union legislator must balance these interests.[51] The Court developed this thinking in the *Balkan-Import-Export* case, which concerned the import of sheep's milk cheese into Germany. Under a regulation the Member States' authorities could, under certain circumstances, levy a countervailing charge on imports, in order to counteract exchange-rate fluctuations. An importer challenged the lawfulness of the regulation and among other things argued that the regulation prioritized the interests of Union agricultural producers at the expense of consumers, and that this was contrary to Article 39. The Court of Justice stated that Article 39 lists various objectives for the common agricultural policy, and that:

> in pursuing these objectives, the community institutions must secure the perma-
> nent harmonisation made necessary by any conflicts between these aims taken
> individually and, where necessary, allow any one of them temporary priority in
> order to satisfy the demands of the economic factors or conditions in view of which
> their decisions are made.[52]

The *Balkan-Import-Export* doctrine has been followed in a number of subsequent cases.[53] In the *Ludwigshafener Walzmühle* case, the Court added to the *Balkan-Import-Export* doctrine that, in reconciling the various objectives of the common agricultural policy, the Union's institutions must avoid putting emphasis on one of these objectives in such a way as to render impossible the realization of other objectives.[54]

---

49. Joined Cases 106 and 107/63 *Toepfer v. Commission* ECR English special edition 405. See also Case 34/62 *Germany v. Commission* ECR English special edition 131.
50. Kaiser, *Grindriß des Agrarwirtschaftsrechts der Europäischen Union 26* (Universität für Boden-kultur, Wien,1996) stated: '[L]ange Zeit ist aus politischen und wirtschaftlichen Gründen den landwirtschaftlichen Interessen gegenüber den Verbraucherinteressen der Vorrang eingeräumt worden'.
51. Case 5/67 *Beus* ECR English special edition 125.
52. Case 5/73 *Balkan-Import-Export* [1973] ECR 1091, para. 24.
53. See most recently Joined Cases C-133/93, C-300/93 and C-362/93 *Crispoltoni* [1994] ECR I-4863, para. 32.
54. Joined Cases 197 to 200, 243, 245 and 247/80 *Ludwigshafener Walzmühle* [1981] ECR 3211, para 41. Compare with the Court of Justice's interpretation of Art. 2 of Council Regulation (EC) No 1184/2006 applying certain rules of competition to the production of, and trade in,

Where the TFEU gives Union institutions discretionary powers, the Court of Justice's examination of the exercise of these powers involves examining whether a decision is clearly a manifest error or an expression of a misuse of power or whether the institution in question has exceeded the scope of its discretion. In his Opinion in the *Martini* case, Advocate General Geelhoed pointed out that, in the context of an examination of the exercise of discretion, the criterion of manifest error or misuse of power means that this power must be exercised for the sake of the objectives of the common agricultural policy.[55] In the *Bela-Mühle* case, the Court of Justice emphasized that although Article 39 enables the common agricultural policy to be defined in terms of a wide choice of measures involving guidance or intervention, the fact nevertheless remains that the second subparagraph of Article 40(3) provides that the common organization of the agricultural markets must be limited to pursuit of the objectives set out in Article 39. Thus, even though the institutions may have wide discretion in implementing measures for the common agricultural policy, this discretion is not unlimited.[56] In the *Roquette Frères* case, the Court of Justice stated generally that when implementing the Union's agricultural policy, the legislator has to evaluate a complex economic situation, and the discretion of the legislator does not apply exclusively to the nature and scope of the measures to be taken but also to some extent to the finding of the basic facts, where the legislator has to make an overall assessment. It thus concluded that in reviewing the exercise of discretion the Court must confine itself to examining whether it contains a manifest error or constitutes a misuse of power or whether the exercise of discretion clearly exceeded its bounds.[57]

[B]   The Individual Objectives

[1]   To Increase Agricultural Productivity (Article 39(1)(a) TFEU)

The statement in Article 39(1)(a) that one of the objectives of the common agricultural policy is to increase agricultural productivity emphasizes that the policy is concerned with the structure of agriculture in a broad sense. An example of Union legislation which expressly referred to Article 39(1)(a) was Council Regulation (EEC) No 797/85 on improving the efficiency of agricultural structures which, in its first recital stated that it is not possible to achieve the objectives of the common agricultural policy set out

---

agricultural products, in which there appears to be a requirement for the fulfilment of all the objectives in Art. 39 TFEU; see Ch. 3, §3.02[F][2].

55.   Opinion of Advocate General Geelhoed in Case C-228/99 *Martini* [2001] ECR I-8401, point 35.
56.   Case 114/76 *Bela-Mühle* [1977] ECR 1211, para. 6. Barents, *The Agricultural Law of the EC* 39 (Kluwer, 1994), states that Art. 39(1) 'does not constitute a substantial restriction on the discretionary power attributed to the Community', while McMahon, *EU Agricultural Law* 25 (Oxford University Press, 2007), states that 'it is important to conclude that the discretion is not unlimited'; see also McMahon, 'The Common Agricultural Policy: from Quantity to Quality', N. Ir. Leg. Q. 9, 12–14 (2002).
57.   Case 138/79 *Roquette Frères* [1980] ECR 3333, para. 25.

in (now) Article 39 (1)(a) and (b) of the Treaty 'without aiding the improvement of the efficiency of agricultural structures'.[58]

Subparagraph (a) does not set any limits to how productivity is to be increased and therefore leaves it to the Union legislator to decide whether the objective is to be achieved exclusively by public regulation or through private initiative.[59]

The scope of discretion of the Union legislator also means that the opportunities of individual farmers can be restricted.[60] The *Hauer* case concerned Council Regulation (EEC) No 1162/76 on measures designed to adjust wine-growing potential to market requirements. The Regulation prohibited the planting of new varieties of vines for a two-year period. The Court of Justice stated that the Regulation should be considered in the context of the common organization of the market in wine, which is closely linked to the structural policy envisaged by the Union in the area in question. The Court ruled that the structural improvement of the wine sector lay within the scope of the objectives set out in Article 39 TFEU.[61]

[2] To Ensure a Fair Standard of Living for the Agricultural Community (Article 39(1)(b) TFEU)

Article 39(1)(b) TFEU sets out, as the second objective of the common agricultural policy, ensuring a fair standard of living for the agricultural community, in particular by increasing the individual earnings of persons engaged in agriculture.

It has been debated whether 'thus', the first word of subparagraph (b), indicates that the subparagraph is subordinated to the objective of subparagraph (a).[62] In the *Danske Landboforeninger* case, the Court of Justice stated that the very wording of Article 39(1)(b) shows that the increase in individual earnings of persons engaged in agriculture is envisaged as being primarily the result of the structural measures described in subparagraph (a).[63]

---

58. Council Regulation (EEC) No 797/85 on improving the efficiency of agricultural structures (OJ L 93, 30 Mar.1985, 1). This Regulation was repealed by Council Regulation (EC) No 1257/1999 on support for rural development from the European Agricultural Guidance and Guarantee Fund (EAGGF) and amending and repealing certain Regulations (OJ L 160, 26 Jul. 1999, 80). Regulation (EC) No 1257/1999 merely refers to the objectives of the common agricultural policy in Art. 39(1); see the first recital.
59. See Olmi in *Le droit de la CEE* (Commentaire Megret), Vol. 2, 'Politique agricole commune' 15 (Université de Bruxelles, Ed. 2, 1991). In Case C-11/88 *Commission v. Council* [1989] ECR 3799, the Court of Justice annulled a Directive on undesirable substances and products in animal nutrition. The Court stated that although the Directive may refer incidentally to certain products not included in Annex I to the TFEU, it applies essentially to products within that Annex, and that it is an essential factor in increasing agricultural productivity, which is the stated objective of Art. 39(1)(a) TFEU. Consequently, the Directive should have been adopted by the Council on the basis of Art. 43 alone, and since that was not the case, it had to be declared void.
60. Barents, *The Agricultural Law of the EC* 34 (Kluwer, 1994).
61. Case 44/79 *Liselotte Hauer* [1979] ECR 3727, paras. 24–27.
62. The French language version of the TFEU uses *ainsi*, and the German version uses *auf diese Weise*.
63. Case 297/82 *Danske Landboforeninger* [1983] ECR 3299, para. 8. Barents, *The Agricultural Law of the EC* 34–35 (Kluwer, 1994), is critical of this and has argued that subparagraph (b)

In practice, the statement of the objective in subparagraph (a) has prompted the question of whether the Union legislator can reduce guaranteed prices for agricultural products.

Some have considered whether, at least in the longer term, subparagraph (b) implies a guaranteed income.[64] However, this view has not received support either from the decisions of the Court of Justice or Union legislation.

The Court of Justice has expressly rejected the idea that subparagraph (b) should be construed as guaranteeing processing industries a certain profit margin.[65]

In some cases, the legislator has maintained guaranteed prices at the same time as introducing levies which effectively reduce prices. Council Regulation (EEC) No 1079/77 on a co-responsibility levy and on measures for expanding the markets in milk and milk products introduced a levy which had the consequence of reducing prices for most milk producers. In the *Stölting* case, the Court ruled that, since the Regulation was intended to contribute to the stabilization of the markets in question, it remained within the limits laid down by Articles 39 and 40.[66] As stated above in Chapter 1, §1.01, the Agenda 2000 plan involved a reduction of price supports for several market organizations, including a 15% reduction for the organizations for arable products, milk and milk products.

[3]   *The Stabilization of Markets (Article 39(1)(c) TFEU)*

The statement, in Article 39(1)(c), that one of the objectives of the common agricultural policy is to stabilize markets has not played a significant role in practice.[67] Since the provisions refers to 'markets', it can authorize regulations on pricing, income support and production quotas.[68] Subparagraph (c) does not require the Union legislator to take a conservative approach.[69] The legislator has no obligation to preserve established rights, and individual producers cannot claim that existing rights or market conditions should remain unchanged by reference to subparagraph (c). This was established in the *Werhahn Hansamühle* case where the Court of Justice stated that the

---

requires both short-term and longer-term policies, and a weighing of these two elements 'constitutes a discretionary matter for the institutions'.

64.   See McMahon, *EU Agricultural Law*, 22 (Oxford University Press, 2007).
65.   Case 281/84 *Zuckerfabrik Bedburg* [1987] ECR 49, para. 23.
66.   Case 138/78 *Stölting* [1979] ECR 713, paras. 4–5.
67.   Usher, *EC Agricultural Law* 37 (Oxford University Press, ed. 2, 2001), merely says: 'The aim of stabilizing markets must in reality be read with the aim of assuring the availability of supplies'.
68.   Case 46/86 *Romkes* [1987] ECR 2671, concerned the regulation of fisheries, and in paragraph 22 the Court of Justice stated that restricting the quantities of fish which may be caught in the short term, and fixing fishing quotas, enables certain species of fish to be conserved and thus contributes to the stabilization of the markets in the long term. On the relationship between Art. 39(1)© TFEU and the levy system in the milk production sector, see Case C-34/08 *Azienda agricola Disaro' Antonio e.a.* [2009] ECR I-4023.
69.   Barents, *The Agricultural Law of the EC* 35 (Kluwer, 1994), considers that subparagraph (c) refers to a short-term policy of 'up to one year', but there appears to be little basis for such a restriction of the time frame for the application of the provision.

concept of stabilization of the markets does not mean the maintenance, at all costs, of positions established under previous market conditions.[70]

[4]   *Assuring the Availability of Supplies (Article 39(1)(d) TFEU)*

So far the common agricultural policy has been to some extent based on the presumption that security of supply can only be assured if there is over-production.[71] Security of supply can also lead to a restriction on commercial opportunities.[72] The term 'supplies' not only includes agricultural products, but also all elements of agricultural production. Thus, according to Article 39(1)(d), it does not matter whether a Union legal act is to the advantage of agricultural producers, consumers or processing industries. With reference to subparagraph (d), the Court of Justice has stated that a directive which was intended to make it possible for the pharmaceutical industry to procure an agricultural raw material at reasonable prices and in sufficient quantities did contribute to the objective of the common agricultural policy of assuring the availability of supplies.[73]

[5]   *Ensuring that Supplies Reach Consumers at Reasonable Prices (Article 39(1)(e) TFEU)*

The objective stated in Article 39(1)(e) – to ensure that supplies reach consumers at reasonable prices – refers not only to the end user, but presumably also includes industries that use agricultural products in their production.[74]

    The word 'reasonable' leaves considerable discretion to the Union legislator, but in the *Balkan-Import-Export* case the Court of Justice indicated that Union legal acts that lead to clearly unreasonable consumer prices could be overridden.[75] In another case, Germany had sought the permission of the Commission to suspend part of the Union customs duties applicable to fresh sweet oranges imported from third countries.

---

70. Joined Cases 63 to 69/72 *Werhahn Hansamühle* [1973] ECR 1229, para. 12, and repeated in Case 106/81 *Kind* [1982] ECR 2885, para. 25.
71. Usher, *EC Agricultural Law* 37 (Oxford University Press, 2d ed., 2001), states that over-production is necessary, 'since crops and animals cannot be persuaded to grow exactly to numbers or quantities required by the planners' predictions'. Barents, *The Agricultural Law of the EC* 35 (Kluwer, 1994), points out that supply could also be assured through imports from third countries.
72. In Joined Cases 3, 4 and 6/76 *Kramer* [1976] ECR 1279, paras. 23 and 33, the Court of Justice stated, by reference to Art. 39(1)(d), that the Union had the competence to undertake international obligations with a view to protecting maritime resources.
73. Case C-131/87 *Commission v. Council* [1989] ECR 3743, paras 23–24.
74. Barents, *The Agricultural Law of the EC* 36 (Kluwer, 1994), with reference to Case C-131/87 *Commission v. Council* [1989] ECR 3743, presumably para. 23. However, Olmi in *Le droit de la CEE* (Commentaire Megret), Vol. 2 – 'Politique agricole commune' 19 (A. Pedine, ed. 2, 1991) takes a different view.
75. Case 5/73 *Balkan-Import-Export* [1973] ECR 1091, para. 24. The individual objectives in Art. 39(1) TFEU can be given different priorities at different times; see §2.05[A] above. Cf. Barents, *The Agricultural Law of the EC* 36 (Kluwer, 1994), who finds that: 'Measures which lead to clearly unreasonable prices for the consumer constitute an infringement of this provision'.

The Commission issued a decision rejecting the application. Germany sought to overturn the decision on the ground, among others, that the Commission's decision was contrary to Article 39(1)(e), because it led to an increased price for the oranges. The Court ruled that 'reasonable prices' does not mean the lowest possible prices, and that reasonableness has to be considered in the context of the common agricultural policy.[76]

[C]   The Special Considerations (Article 39(2) TFEU)

In extension of the statement of objectives in Article 39(1) TFEU, Article 39(2) sets out three considerations which the Union institutions must take into account when developing the common agricultural policy.[77]

The first consideration to be taken into account is the particular nature of agricultural activity. According to Article 39(2)(a), agricultural activity has a particular nature resulting from the social structure of agriculture and from structural and natural disparities between the various agricultural regions. Subparagraph (a) has played a special role in justifying the targeting of agricultural legislation at specific agricultural sectors or particular geographic regions of the EU.[78]

In the *Eridania* case, the Court of Justice confirmed the legality of the then market organization for sugar, under which Italian sugar producers were allocated a production quota that was less than the sugar consumption on the home market. The Court did not refer to Article 39(2)(a), but referred instead to the principle of regional specialization, which requires production to take place where it is most economically suitable. The fact that Italy was given a production quota that was less than the sugar consumption on the home market was not an expression of discrimination, but rather a consequence of the requirement that, in a common market characterized by regional specialization, production in the individual Member States must be able to develop independently of the level of consumption in those States.[79]

The second consideration, referred to in Article 39(2)(b), is the need to make appropriate adjustments by degrees. Subparagraph (b) underlines that the Union legislator has wide discretion. The adjustment of agricultural policy must be 'appropriate', and what may be considered 'appropriate' is largely a political question. The provision puts the obligation on the legislator to make adjustments by degrees.[80]

---

76. Case 34/62 *Germany v. Commission* ECR English special edition 131.
77. Usher, *EC Agricultural Law* 39 (Oxford University Press, 2d ed., 2001), finds that 'there would appear to be little difference in status between those "factors to be taken into account" and the objectives of the common agricultural policy strictly so called'.
78. The Court of Justice has stated that the need to treat different groups of those engaged in agriculture differently is recognized in Art. 39(2)(a): see Case 139/77 *Denkavit* [1978] ECR 1317, para. 15; and Case 36/79 *Denkavit* [1979] ECR 3439, para. 16. The first recital of the now repealed Council Directive 75/268/EEC on mountain and hill farming and farming in certain less-favoured areas referred to Art. 39(2)(a).
79. Case 250/84 *Eridania zuccherifici* [1986] ECR 117, paras. 20–21.
80. Barents, *The Agricultural Law of the EC* 39–40 (Kluwer, 1994), takes a different view and considers that the words 'by degrees' means that the Union legislator is free to decide how the adjustment may be made.

Finally, Article 39(2)(c) requires the Union institutions to take into account that agriculture constitutes a sector closely linked with the economy as a whole. In a case in which the United Kingdom challenged the validity of Council Directive 85/649/EEC prohibiting the use in livestock farming of certain substances having a hormonal action, the Court stated that Article 39(2)(c) means that agricultural policy objectives must be conceived in such a manner as to enable the Union institutions to carry out their duties in the light of developments in agriculture and in the economy as a whole.[81] By a reference to subparagraph (c), the Court has established that it is not the aim of the common agricultural policy to exclude agricultural producers from the effects of a national income policy.[82]

---

81. Case 68/86 *United Kingdom v. Council* [1988] ECR 855, para. 10.
82. Case 297/82 *Danske Landboforeninger* [1983] ECR 3299, para. 8.

CHAPTER 3

Legal Basis and Procedure

§3.01 THE REQUIREMENT FOR LEGAL BASIS

The requirement for legal basis plays an important role in Union law. It means that Union legislative acts cannot be adopted without the necessary basis in a Treaty provision. The principle is established in Articles 5(2) and 13(2) of the Treaty on European Union that state that the EU's institutions must act within the limits of the powers conferred on them in the Treaties, and in conformity with the procedures, conditions and objectives set out in the Treaties.

The treaty basis for the adoption of legislation on the common agricultural policy is in Article 43 of the TFEU.[1] Article 43(2) TFEU authorizes the European Parliament and the Council to establish the common organization of agricultural markets, and adopt other provisions necessary for the pursuit of the objectives of the common agricultural policy. Article 43(3) TFEU authorizes the Council to adopt measures on fixing prices, levies, aid and quantitative limitations. Before the Lisbon Treaty entered into force, Article 37 of the Treaty establishing the European Community (EC Treaty) gave express power to the Council to make regulations, issue directives, take decisions, and make recommendations. It must be assumed that the changed wording of Article 43(2) and (3) TFEU does not imply any change to the kinds of legal acts for which the Treaty gives power to adopt in the area of agriculture. The 'provisions' and 'measures' referred to in Article 43(2) and (3) must thus be assumed to referred to the legal acts listed in Article 288 TFEU.

---

1. The scope of the legal basis in the treaties for the individual policy areas in relation to each other raises special questions; see e.g., Hartley, *The Foundations of European Union Law* 118 (Oxford University Press, 7th ed., 2010). Under the doctrine of implied powers, Art. 43(2) and (3) imply that the Union has powers to act under international law, including entering into treaties in the area of agriculture. See Joined Cases 3, 4 and 6/76 *Kramer* [1979] ECR 1279, para. 33, where the Court of Justice ruled that the Union has power to enter into international commitments for the conservation of the resources of the sea.

Legal acts that are adopted by the institutions on the basis of EU legislation must comply with the authorizing legislation.[2] Often, Union legislation in the area of agriculture authorizes the Commission to issue legal acts. Such legislation is not limited to authorizing the Commission to issue decisions, but often gives the Commission competence to issue regulations and directives. In connection with proceedings to annul a Commission Regulation which reserved the use of the term 'feta' as a designation of origin for a particular kind of cheese produced in Greece, the Court of Justice annulled the Regulation to the extent to which it registered 'feta' as a protected designation of origin. The Court held that the Commission Regulation did not comply with the conditions in the Regulation adopted by the Council which empowered the Commission to issue regulations on geographical designations of origin for agricultural products.[3]

In a case against Ireland for breach of the Treaty by an Irish measure prohibiting the import into Ireland of potatoes from third countries, Ireland argued that the Commission had given Ireland the impression that the Irish measures were not objectionable under Union law. On this, the Court of Justice stated that regardless of the circumstances, the Commission cannot confer on a Member State the right to maintain provisions which are objectively contrary to Union law.[4]

In the area of agriculture, the requirement for proper legal basis has raised several problems of principle, and the solutions to these have been significant for the development of Union law in general. The first concerns the scope for discretion which the Treaty allows the Union legislator. When Article 43(2) and (3) TFEU gives legislative competence with regard to the establishment of the common agricultural policy, the legislative power is linked to the interpretation of the statement of objectives in Article 39 TFEU.[5] In the *Bayerische* case, the Court of Justice referred in general to 'one of the chief features is the exercise of a wide discretion essential for the implementation of the common agricultural policy'.[6]

---

2.   If the Council wishes to amend legislation which it has itself adopted, it must carry out a new formal legislative procedure. The Council does not have a general competence to amend legislation merely by means of a Council decision.
3.   Joined Cases C-289/96, C-293/96 and C-299/96 *Denmark v. Commission* [1999] ECR I-1541. Provisions on the protection of geographical indications and designations of origin for agricultural products and foodstuffs are now in Council Regulation (EC) No 510/2006 (OJ L 93, 31 Mar. 2006, 12).
4.   Case 288/83 *Commission v. Ireland* [2985] ECR 1761, para. 22.
5.   On the interpretation of Art. 39, see above in Ch. 2, §2.05.
6.   Joined Cases 83 and 94/76, 4, 15 and 40/77 *Bayerische* [1978] ECR 1209, para. 6. The Court of Justice has subsequently expressly referred to paragraph 6 in the *Bayerische* case in, among others, Case C-390/95 P *Antillean Rice Mills* [1999] ECR I-769, para. 57. In Joined Cases 197 to 200, 243, 245 and 247/80 *Ludwigshafener Walzmuehle* [1981} ECR 3211, para. 37, the Court of Justice stated that in determining their policy in the area of agriculture, the competent Union institutions enjoy wide discretionary powers regarding not only the establishment of the factual basis of their actions but also the definition of the objectives to be pursued within the framework of the provisions of the Treaty and the choice of the appropriate means of action. Barents, *The Agricultural Law of the EC* 54 (Kluwer, 1994), stated: 'The power of Article [43] is …a political power'. On judicial review of the Council's exercise of its power pursuant to Art. 37 of the EC Treaty, see Barents, 'Recent Developments in Community Case Law in the Field of Agriculture', Com. Mkt. L. Rev. 811, 817–818 (1997).

The wide scope of the Union legislator's discretion based on Article 43(2) and (3) TFEU involves both an assessment of whether legislation is necessary and, if so, the content of the legislation.[7] Where the legislator delegates competence to the Commission, the scope of the Commission's discretion is determined by the delegating legal act.[8]

In the *Jippes* case the Court of Justice summarized its practice on the judicial review of the legislator's exercise of discretion pursuant to Article 43 TFEU:

> the Community legislature enjoys a wide discretionary power in matters concerning the common agricultural policy, corresponding to the political responsibilities given to it by Articles [40-43 TFEU]. Consequently, judicial review must be limited to verifying that the measure in question is not vitiated by any manifest error or misuse of powers and that the authority concerned has not manifestly exceeded the limits of its power of assessment.[9]

In practice, this means that the Court only reviews whether the Union legislator has acted arbitrarily.[10]

Another problem of principle connected with the requirement for legal basis is the delimitation of the scope of the power in Article 43(2) and (3) TFEU in relation to other powers to legislate in the Treaty. If EU Institutions adopts a measure on the basis of the power given in a Treaty provision, but the measure should have been based on some other provision, the situation is characterized as a misuse of power. This was established in one of the earlier judgments of the Court of Justice.[11] The Court has also emphasized that the question of what is the right basis for a legal act is not a matter for the discretion of the legislator, but must be decided on an objective basis which the Court can review.[12]

Finally, the requirement for legal basis has meant that a formal question is raised before the Court of Justice. In the *Roquette* case, before to the entry into force of the Lisbon Treaty, the Council had adopted a Regulation on the basis of Article 37 EC, setting production quotas for isoglucose. The Council did consult the European Parliament in accordance with the requirements of Article 37 EC but did not await a reply from the Parliament before adopting the regulation. The Court ruled that the

---

7. Case 113/88 *Leukhardt* [1989] ECR 1991, para. 20, where the Court stated that it has consistently held that when a situation requires the evaluation of a complex economic situation, as with the common agricultural policy, the Community legislature enjoys a wide discretion as to the nature and scope of the measures to be taken.
8. See Case 78/74 *Deuka* [1975] ECR 421, para. 8.
9. Case C-189/01 *Jippes* [2001] ECR I-5689, para. 80, where the Court referred to Case C-331/88 *Fedesa and Others* [1990] ECR I-4023, paras 8 and 14, which is based in turn on Case 138/79 *Roquette* [1980] ECR 3333, para. 25. The Court repeated para. 80 in the *Jippes* case in Case C-310/04, *Spain v. Council* [2006] ECR I-7285, para. 96.
10. Barents, *The Agricultural Law of the EC* 57 (Kluwer, 1994. Hartley, *The Foundations of European Union Law* 426 (Oxford University Press, 7th ed., 2010), points out that misuse of powers is more difficult to prove than other grounds for invalidity.
11. Case 15/57 *Hautes Fournaux*, ECR English special edition 211.
12. Case 131/86 *United Kingdom v. Council* [1988] ECR 905, para. 29. In the case the Court rejected a claim that a directive on battery hens was contrary to Art. 253 EC (now Art. 296 TFEU) on the duty to state the reasons on which a legal act is based, as it as its legal basis only referred to Art. 37 EC (now Art. 43 TFEU).

requirement of consultation 'implies that the Parliament has expressed its opinion', and disregarding this requirement meant that the measure was void.[13]

§3.02   LIMITS TO THE UNION LEGISLATOR'S DISCRETION

The Court of Justice's review of the discretion exercised by the Union legislator pursuant to Article 43 TFEU is largely based on a number of the basic principles of Union law, but Union fundamental rights also play a role. However, the role of the basic principles and fundamental rights is not merely to set limits to the legislator's discretion, as they also play an important role in the interpretation of Union legislation.[14]

[A]   The Doctrine of Misuse of Power

The Union legislator's discretion is limited by the prohibition of the misuse of power. Misuse of power has been defined by the Court as 'the adoption by a Community institution of a measure with the exclusive or main purpose of achieving an end other than that stated or evading a procedure specifically prescribed by the Treaty for dealing with the circumstances of the case'.[15]

The prohibition of the misuse of power does not mean that the institutions cannot take account of political considerations when exercising their discretion.[16]

The application of the misuse of powers doctrine in the area of agriculture is illustrated by the *Compagnie d'approvisionnement* case. The case concerned two regulations which the Council had adopted in order to counter the effects of the devaluation of the French franc, among other things by giving support to French exporters of agricultural products. However, a French exporter argued that the support was insufficient. The Regulation had been adopted on the basis of Article 103 of the EEC Treaty which gave the Council wide discretionary powers to 'decide upon the measures appropriate to the situation'. The Court of Justice did not accept the argument of the exporter, because Article 103 EEC empowered the Council to adopt measures 'without obliging it to do so' and 'Article 103 conferred on that institution a wide power of discretion to be exercised in accordance with the "common interest" and not with the individual interests of a specific group of traders'.[17]

---

13. Case 138/79 *Roquette* [1980] ECR 3333, para. 33. On the present legislative procedure, see below in §3.05.
14. See §3.03 below on the influence of the basic principles on the interpretation of Union legislation. Sometimes a principle can be absolute and failure to comply with the principle can affect the validity of legislation, and sometimes principles are used as an aid to the interpretation of legislation.
15. Case C-156/93 *European Parliament v. Commission* [1995] ECR I-2019, para. 31, on the validity of legislation on the organic production of agricultural products.
16. Case 57/72 *Westzucker* [1973] ECR 321, para. 17, where the Commission may have given in to political pressure from France and Italy.
17. Joined Cases 9 and 11/71 *Compagnie d'approvisionnement* [1972] ECR 391, paragraph 33.

[B] The Principle of Equal Treatment

Union law contains a general principle of equal treatment, according to which similar situations may not be treated differently unless there is an objective justification for doing so.[18] This principle is particularly emphasized in a number of provision in the TFEU: Article 18 prohibits discrimination on grounds of nationality; Article 157 establishes the principle of equal pay for male and female workers for equal work or work of equal value; and Article 40(2), second paragraph, excludes any discrimination between producers or consumers within the Union in the area of agriculture.

The *Ruckdeschel* case concerned the validity of a Regulation on production subsidies for maize. The Regulation distinguished between maize used for the production of maize flour (*quellmehl*), and maize used for the production of starch. According to the Regulation, production subsidies could only be given for the production of starch and not for the production of maize flour. The Court of Justice stated, with reference to Article 40(2), second paragraph, that:

> the prohibition of discrimination laid down in the aforesaid provision is merely a specific enunciation of the general principle of equal treatment which is one of the fundamental principles of Community law. This principle requires that similar situations shall not be treated differently unless differentiation is objectively justified.[19]

In the above case the Court of Justice subjected the facts of the case to a characteristic review in relation to the principle of equal treatment. First the Court considered whether the producers of maize flour and starch were in similar situations. When the Court found this to be the case, it considered whether the producers had been treated differently in the legislation in question. As this was clearly the case, the Court finally considered whether this different treatment was objectively justified.[20] The burden of proving that the difference in treatment if objectively justified lies on the Union legislator. In this case there was a question of whether the maize flour could be used for animal feed, and the Court found that, during the proceedings, the Commission had been asked by the Court to show evidence that maize flour could have such a use, but it had not been able to provide such evidence.[21]

---

18. Joined Cases 117/76 and 16/77 *Ruckdeschel* [1977] ECR 1753, paragraph 7; and Joined Cases 124/76 and 20/77 *Pont-à-Mousson* [1977] ECR 1795, paragraph 17.
19. Joined Cases 117/76 and 16/77 *Ruckdeschel* [1977] ECR 1753, paragraphs 7 and 10. See also Case 280/93 *Germany v. Council* [1994] ECR I-4973, paragraph 67.
20. On the Court's customary application of the principle of equal treatment, see Usher, *EC Agricultural Law* 42 (Oxford University Press, 2d ed., 2001).
21. Joined Cases 117/76 and 16/77 *Ruckdeschel* [1977] ECR 1753, paragraph 9. With reference to Joined Cases 103 and 145/77 *Royal Scholten-Honig* [1978] ECR 2037, Usher, *EC Agricultural Law* 42 (Oxford University Press, 2d ed., 2001), cautiously concluded that: 'the approach of the Court ... appears to have been that it was for the Community institutions to show that the difference was objectively justified rather than for the producers affected to show that it was not justified'. With reference to the same judgment, McMahon, *EU Agricultural Law* 29 (Oxford University Press, 2007), stated that '[t]he burden of proof is on the Community institutions to show the existence of objective justification'.

Article 40(2), second paragraph, concerns the equal treatment of producers and consumers, but the general principle of equal treatment also covers other groups of economic operators, such as exporters and importers. In a case concerning the common organization of the market for bananas, the Court of Justice found that the organization in question introduced rules for economic operators who are neither producers nor consumers. However, the Court ruled that because of the general nature of the principle of equal treatment, the prohibition of discrimination also applied to other categories of economic operators who are subject to a common organization of a market.[22]

[C] The Principle of Proportionality

The first time that the Court of Justice established that the proportionality principle is a part of Union law was in a case concerning agricultural law.[23] The principle was subsequently expressly referred to in Article 5 of the EC Treaty, and today it is expressed in Article 5(4) of the Treaty on European Union (EU Treaty). According to the principle, the content and form of legal acts of the EU institutions may not exceed what is necessary and appropriate to achieve the objectives of the Treaties. This means that if it is possible to choose between two or more suitable measures, the least burdensome measure must be chosen, and the burdens of a measure may not be disproportionate in relation to its aims.[24]

As for judicial review of the conditions for applying the proportionality principle in the context of the wide discretion which the Union legislator has in relation to agricultural policy, the legality of a measure adopted in the sphere of this policy area can only be affected if the measure is manifestly inappropriate, given the objective pursued by the measure.[25] In applying the proportionality principle, it is not for the Court of Justice to examine whether a measure adopted by the legislature was the only one or the best one possible, but whether it was manifestly inappropriate.[26]

---

22. Case C-280/93 *Germany v. Council* [1994] ECR I-4973, para. 68.
23. Case 11/70 *Internationale Handelsgesellschaft* [1970] ECR 1125, paragraph 16.
24. Case C-189/01 *Jippes* [2001] ECR I-5689, para. 81, where the judgment is based on Case C-331/88 *Fedesa and others* [1990] ECR I-4023, para. 13; and Joined Cases C-133/93, C-300/93 and C-362/93 *Crispoltoni* [1994] ECR I-4863, para. 41. The Opinion of Advocate General Dutheillet de Lamothes in Case 11/70 *Internationale Handelsgesellschaft* [1970] ECR 1125, is often referred to as introducing the proportionality principle into Union law. McMahon, *EU Agricultural law* 31 (Oxford University Press, 2007), refers to the proportionality principle and the principle of the protection of legitimate expectations, as the general principles that are most relevant for private individuals in connection with the rules of the common agricultural policy. See also Leidwein, *Einführung in das europäisches Agrarecht* 33-34 ((Berliner Wissenschafts-verlag, Österreichischer Agrarverlag and Neuer wissenschaftlicher Verlag,1996). The applica-tion of the force majeure doctrine, in connection with a requirement for economic operators to provide guarantees, can be regarded as a concrete example of the proportionality principle; see Case 25-70 *Köster* [1970] ECR 1161, para. 22. On force majeure in connection with the provision of guarantees for import and export licenses and similar permits, see below in Ch. 4, §4.06[C].
25. Case C-189/01 *Jippes* [2001] ECR I-5689, para. 82, and the cases referred to there.
26. Case C-310/04 *Spain v. Council* [2006] ECR I-7285, para. 99, referring 'to that effect' to Case C-189/01 *Jippes* [2001] ECR I-5689, para. 83.

The Court's application of the proportionality principle in connection with the judicial review of Union legislation on agriculture is illustrated by the *Bela-Mühle* case. In 1976, the Council adopted a Regulation which obliged certain agricultural producers to buy skimmed milk powder. By this means, the Council sought to reduce the very large stocks of skimmed milk powder in the EU; these had been built up as a consequence of purchases by intervention agencies under the common organization of the market for milk and milk products. The Regulation compelled producers of concentrated animal feed to use skimmed milk powder in their production instead of soya oilcake. The problem for such producers was that skimmed milk powder was about three times more expensive than soya oilcake. The Regulation prompted the bringing of several cases before national courts which referred such cases to the Court of Justice for a preliminary ruling on whether the Regulation was invalid under the proportionality principle.

The Court of Justice ruled that the Regulation was invalid. The obligatory purchase at such a disproportionately high price was a discriminatory allocation of the burden between different agricultural sectors, 'nor … was such an obligation necessary in order to attain the objective in view, namely, the disposal of stocks of skimmed-milk powder.'[27]

[D]   The Protection of Legitimate Expectations

The principle of the protection of legitimate expectations requires that a Union legal act does not breach the legitimate expectations of persons to whom it is addressed unless the restriction of rights if justified by the public interest. In its decisions, the Court of Justice has emphasized, a little pedagogically, that if a prudent and circumspect trader could have foreseen that the adoption of a Union measure would be likely to affect his interests, he cannot plead the protection of legitimate expectations if the measure is adopted. While the principle of the protection of legitimate expectations is a fundamental principle of Union law, traders cannot have a legitimate expectation that an existing situation which is capable of being altered by EU institutions in the exercise of their discretion will remain unchanged.[28]

The principle of the protection of legitimate expectations is relatively frequently relied on in cases against the EU institutions, but seldom successfully. Speculative attempts to benefit from support schemes cannot expect to find support from the principle.[29] This is illustrated by the *Mackprang* case. The case concerned the common

---

27. Case 114/76 *Bela-Mühle* [1977] ECR 1211, paragraph 7.
28. Joined Cases C-37/02 and C-38/02 *Di Lenardo Adriano* [2004] ECR I-6911, paragraph 70, and the cases referred to there.
29. Examples of cases in which the Court of Justice has held that Union legal acts in the form of a regulation have been contrary to the principle of the protection of legitimate expectations include: Case 120/86 *Mulder* [1988] ECR 232, paras 24–27; Case 170/86 *Deetzen* [1988] ECR 2355, paras 13–16; Case C-189/89 *Spagl* [1990] ECR I-4539, para. 29; and Case C-217/89 *Pastätter* [1990] ECR I-4585, paragraph 20. All these cases concerned an additional levy on milk. On some of the milk cases in the light of the principle of the protection of legitimate expectations, see Thiele, *Agrarrecht* 333 (1988). In Case T-310/06 *Hungary v. Commission* [2007] ECR II-4619, paras 62–72, the Court invalidated a regulation which introduced a new

organization for the market in cereals which, at the time of the case, meant that intervention agencies were obliged to buy cereals at intervention prices. Cereals could be offered to the intervention agencies in the various Member States and, in principle, there was no advantage for the seller to choose an intervention agency in one Member State rather than another. The usual procedure was for the seller to offer their cereals to the intervention agency of the Member State where they were produced. However, in 1969 the value of the French franc fell, and it became profitable for German corn merchants to buy cereals in France and sell them to the German intervention agency. This practice threatened to exceed the storage capacity of the German agency and otherwise to bring the German intervention agency to its knees. The Commission therefore issued a decision authorizing the German agency to restrict its purchases to wheat and barley grown in Germany. The decision was issued on 8 May and entered into force on the same day. The decision expressly stated that it did not apply to cereals which had already been offered to the German intervention agency before the decision's entry into force.[30]

Mackprang was a German corn merchant who had bought cereals in France with a view to importing them to Germany and selling them to the German intervention agency. On 8 May, it had wheat in transit from France to Germany, which meant that Mackprang was not then entitled to offer it to the German intervention agency. When the wheat arrived in Germany the intervention agency refused to buy it, whereafter Mackprang brought a claim for compensation against the agency in the German courts. Mackprang argued that when it had bought the wheat in France it had a legitimate expectation that it would be able to sell it to the German intervention agency. The German court referred the case to the Court of Justice for a preliminary ruling, and the Court of Justice ruled that the Commission's decision was in line with the aims pursued by the intervention system for cereals, and that the decision did not constitute a breach of the plaintiff's legitimate expectation. The Commission's decision was a justified measure taken against speculative transactions.[31] In his Opinion on the case, Advocate General Warner argued that:

> No trader who was exploiting that situation in order to make out of the system profits that the system was never designed to bestow on him could legitimately rely on the persistence of the situation. On the contrary, the only reasonable expectation that such a trader could have was that the competent authorities would act as swiftly as possible to bring the situation to an end. Nor ... could he expect particular tenderness at their hands.[32]

This view has been become fixed in the case law of the Court of Justice.[33]

---

quality requirement for maize which it was impossible for farmers who had already sowed their crops to comply with. The Court ruled that the Commission had breached the legitimate expectations of these farmers.

30. Case 2/75 *Mackprang* [1975] ECR 607.
31. *Ibid.*, para. 4.
32. *Ibid.* Opinion of Mr Advocate General Warner, 623.
33. See Case C-310/04 *Spain v. Council* [2006] ECR I-7285, para. 81, and the cases referred to there.

The General Court has stated in its case law that it is only possible to rely on the principle of the protection of legitimate expectations if the institution concerned has given 'precise assurances'.[34] The concrete assessment of whether the necessary assurances have been given by an institution is illustrated by the *Lefebvre frères et soeurs* case. A French banana importer claimed to have a legitimate expectation on the basis of two letters from the Commission stating that the Commission would take account of the importer's interests in connection with the adoption of some legislative measures. In one letter, Mr Andriessen, the Vice-President of the Commission, stated that:

> As regards the problems more closely connected with the application of Article 115 of the Treaty, I am pleased that the operators whom you represent are conscious of the fact that the Commission's decisions in that sphere have always taken their concerns into account. I can assure you that if the French authorities request an extension of the measures in force beyond 30 June 1991, the Commission will certainly assess such a request bearing in mind the wishes expressed by you on behalf of your clients.

In the second letter, a representative for the Commission had expressed his 'wish to reassure (the applicants) that, in formulating the proposal to the Council for the establishment of a Community system in the banana sector, the Commission will certainly take into account the particular situation of small and medium-sized importers'.[35]

In its judgment, the General Court made it clear that there is an important difference between a statement made by the Commission in general terms, which cannot engender any valid expectations, and an assurance in precise terms on which expectations may legitimately be based. The statements of the Commission in the letters relied on by the plaintiff fell within the first category, since they were worded in very general terms. Thus those statements were not capable of engendering any valid expectations on the part of the applicants, and the Court rejected the claim that there had been a breach of the principle of the protection of legitimate expectations.[36]

[E]   Fundamental Rights

As in other policy areas, in the agricultural area the Union legislator's discretion is limited by certain fundamental rights in Union law. The term 'fundamental rights' is generally used in respect of freedoms and human rights that are set out in a constitutional context. In the following, such rights are referred to as 'classic fundamental rights'. However, the Court of Justice has also characterized certain other Treaty-based rights as fundamental rights; these rights are here referred to as 'special

---

34. Case T-20/91 *Holtbecker* [1992] ECR II-2599, para. 53. McMahon, *EU Agricultural law*, 31 (Oxford University Press, 2007) states that '[i]n the absence of specific assurances by the Community institutions, it appears unlikely that individuals can claim a breach of the principle of legitimate expectations'.
35. Case T-571/93 *Lefebvre frères et soeurs* [1995] ECR II-2379, paragraph 73.
36. *Ibid.*, paras 72 and 74–75.

fundamental rights under Union law', and in the same way as the classic fundamental rights they set limits to the exercise of discretion by the Union legislator.

[1]   Special Fundamental Rights under Union Law

In a case brought against France for breach of the Treaty, where French agricultural organizations had prevented the import of agricultural products from other Member States, the Court of Justice pointed out that the internal market is an area without internal frontiers in which the free movement of goods is ensured in accordance with the provisions of the Treaty, stating that this 'fundamental principle' is implemented by Article 34 TFEU.[37] In more recent cases, the Court has referred to the freedom of movement of goods as 'a fundamental right conferred by the Treaty'.[38] It must be assumed that the same characteristic applies to the other freedoms, such as the free movement of workers and the freedom to provide services, as well as the free movement of capital.[39]

The special fundamental rights under Union law are expressed in provisions of the TFEU and can only be amended pursuant to Article 48 of the EU Treaty. The Union legislator cannot alter these Treaty provisions, and to the extent that Union legislative acts breach the special fundamental rights under Union law without this being authorized by the Treaty, such legislation must be found invalid.[40]

Within the area of agriculture, the special fundamental rights under Union law must viewed in the context of Article 39(1) and (2) TFEU, which state that the rules laid down for the establishment and functioning of the internal market shall apply to agricultural products unless otherwise provided in Articles 39–44. The special fundamental rights under Union law are the foundations of the internal market, and basis in the Treaty for making exceptions to these rules can be found either in provisions on agriculture in Articles 39–44 TFEU, or in the general rules on exceptions associated with the individual rights, such as the exceptions to the free movement of goods in Article 36 TFEU. To the extent that Union legislation is authorized by Articles 39–44, this legal basis in the Treaty can be characterized as an exception to the special fundamental rights under Union law, including the right of the free movement of goods. It would thus be tautologous, within the area of the common agricultural policy, to treat systematically the rights of free movement as special fundamental rights under

---

37. Case C-265/95 *Commission v. France* [1997] ECR I-6959, para. 27.
38. Case C-228/98 *Dounias* [2000] ECR I-577, para. 64, with para. 63. The Court of Justice has similarly characterized the rules on the free movement of goods in Case C-394/97, *Heinonen*, [1999] ECR I-3599, para. 38, as 'the fundamental freedom', and in Case C-443/98, *Unilever Italia*, [2000] ECR I-7535, para. 40, as 'one of the foundations of the Community'.
39. See Case C-49/89 *Corsica Ferries France* [1989] ECR I-4441, paragraph 8, where the Court held that the Treaty provisions concerning the free movement of goods, persons, services and capital are 'fundamental Community provisions'.
40. In Case C-221/89 *Factortame II* [1991] ECR I-3905, paras 40–42, the Court of Justice rejected the argument that the TFEU rules on the common fisheries policy, with a system of national fishing quotas, could authorize conditions relating to nationality laid down in national law which restricted the right of establishment laid down in Art. 43 of the EC Treaty, now Art. 49 TFEU.

Union law. In the area of agriculture, the rules on the free movement of goods play their part through Article 38(2) TFEU.[41]

[2] Classic Fundamental Rights

Within the area of the common agricultural policy, the classic fundamental rights, such as the freedom to pursue a trade or profession and the protection of property rights, have been considered by the Court of Justice in a number of cases.[42]

In the *Hauer* case, a German wine-grower claimed both these rights.[43] In 1975, Hauer asked the German authorities for permission to plant vines on a parcel of land. This permission was refused, and when the case was referred to the Court of Justice for a preliminary ruling the reason given for this was based on a Regulation which was aimed at reducing the excessive production of wine.[44] The Regulation contained a comprehensive prohibition of the new planting of vines and its Article 2(1) stated: 'All new planting of wine varieties classified as wine grape varieties ... shall be prohibited during the period from 1 December 1976 to 30 November 1978'. For the Court of Justice the question was whether the provision breached Hauer's property rights and freedom to pursue a trade or profession.

The Court of Justice stated that the right to property is guaranteed in the Union legal order in accordance with the ideas common to the constitutions of the Member States, but that the exercise of such rights is subject to the general interest. The Court noted that all the wine-producing countries of the Union had restrictive legislation, of differing severity, concerning the planting of vines, the selection of varieties and the methods of cultivation, and that in none of the countries concerned were those provisions considered to be incompatible in principle with the right to property.[45] The Court found that the Regulation should be considered in the light of legislation that restricted property rights in the general interest, and assessed on that basis, and whether the restrictions it introduced were compatible with the objectives of the common agricultural policy, as set out in Article 39 TFEU. It pointed out that the Regulation had been adopted against the background of the wine harvest in 1974, which was characterized by there being persistent excess production and that according to its wording the Regulation performed a double function. On the one hand it had to enable an immediate brake to be put on the continued increase in the surpluses, and on the other hand it had given the Union institutions the necessary time for implementing a structural policy to encourage high-quality production. The Regulation's

---

41. See §3.02[F] below on Art. 38(2) and the rules on freedom of movement in the area of agricultural law.
42. On publication of information on beneficiaries of agricultural aid, see Joined Cases C-92/09 and C-93/09 *Volker und Markus Schecke GbR* [2010] ECR I-11063.
43. Case 44/79 *Hauer* [1979] ECR 3727. The Court of Justice had established in Case 4/73 *Nold* [1974] ECR 491, para. 12–13, that right to the free pursuit of business activity is a fundamental right ensured by the Court.
44. See Council Regulation (EEC) No 1162/76 on measures designed to adjust wine-growing potential to market requirements.
45. Case 44/79 *Hauer* [1979] ECR 3727, paras 17 and 20–22.

prohibition was only an interim measure aimed at the immediate reduction of surpluses, and at the same time it prepared for more permanent structural measures. Seen in this light, the measure did not entail an undue limitation upon the exercise of the right to property. The restriction on the use of property by the prohibition of the new planting of vines for a limited period was justified by the objectives of the general interest pursued by the Union and did not infringe the substance of the right to property as recognized and protected in the Union legal order.[46]

As for the freedom to pursue a trade or profession, in the *Hauer* case the Court of Justice stated that while this is a special fundamental right under Union law, it should not be treated as an absolute right but rather should be seen in the context of the social function of the right.[47] The prohibition in the Regulation of the new planting of wines did not affect access to the occupation of wine-growing or the freedom to pursue that occupation on land where wines were already being cultivated at the time in question. To the extent that the prohibition of new planting affected the free pursuit of the occupation of wine-growing, it was no more than the consequence of the restriction on the exercise of the right to property, so the two restrictions merged.

Thus any restriction on the free pursuit of the occupation of wine-growing was justified by the same reasons that justified the restriction placed upon the use of property.[48]

The Court of Justice did not decide whether Hauer was entitled to compensation as a consequence of the restriction of her rights by the Regulation, but the Advocate General considered this. He argued that there was not a basis for awarding compensation, both because the restriction was for a limited period and because the intensity of the intervention was very limited.[49]

[F]   The TFEU Rules of the Establishment of the Internal Market in the Area of Agriculture

Article 38(1) TFEU states that the internal market extends to agriculture and trade in agricultural products. As mentioned above in section §3.02[E][1], Article 38(2) TFEU states that the Treaty rules on the establishment and functioning of the internal market

---

46. *Ibid.* paras 23–30. The Court's weighing of the prohibition of new planting against property rights was the same as that suggested in the Opinion of Advocate General Capotorti on the case, point 9. Further light is cast on the protection of property rights (in the area of milk production) in Case 5/88 *Wachauf* [1989] ECR 2609; Case C-2/92 *Bostock* [1994] ECR I-35; and Case C-63/93 *Duff* [1996] ECR I-569. On cases concerning the protection of property rights in the area of milk production, see Usher, *EC Agricultural Law* 50–52 (Oxford University Press, 2d ed., 2001).
47. The Court of Justice had previously established this in Case 4/73 *Nold* [1974] ECR 491, para. 14.
48. Case 44/79 *Hauer* [1979] ECR 3727, paras 31 and 32.
49. Opinion of Advocate General Capotorti in Case 44/79 *Hauer* [1979] ECR 3727, point 8. Usher, *EC Agricultural Law* 49–50 (Oxford University Press, 2d ed., 2001), points out that Art. 8 of Council Regulation (EC) No 1493/1999 on the common organization of the market in wine (OJ L 179, 14 Jul. 1999, 1) allowed for the payment of a premium for the cessation of wine-growing on specified areas. See Council Regulation (EC) No 1234/2007 establishing a common organization of agricultural markets and on specific provisions for certain agricultural products on grubbing-up premiums for years 2010–2011 in connection with wine production.

apply to agricultural products, unless otherwise provided in Articles 39–44.[50] Article 38(1) and (2) concerns the extent to which the Treaty rules on free movement, the competition rules and the rules on State aid apply in the area of agriculture, and where these Treaty rules apply, they restrict the powers of the Union legislator.[51]

Article 38(1) and (2) TFEU mean that the Treaty rules on the free movement of workers and services and freedom of establishment apply without restriction to agriculture, while the rules on the free movement of goods only apply with certain reservations.[52]

The Court of Justice considered the interpretation of Article 38(2) in the *Ramel* case. The case concerned the free movement of goods and in particular the Treaty-based prohibition of customs duties on imports and exports between Member States and of all charges having equivalent effect and an exception to this laid down in a Regulation. At that time, the common organization of the market for wine was regulated by, among others, Regulation (EEC) No 816/70, whose Article 31(1) stated that in the internal trade of the Union, customs duties on imports and exports and of all charges having equivalent effect were prohibited. This provision merely repeated what already applied pursuant to Articles 12 and 30 of the EEC Treaty, now Articles 30 and 34 TFEU. As an exception to this, Article 31(2) of the Regulation provided that:

> *By way of derogation from the provisions of paragraph 1*, so long as all the administrative mechanisms necessary for the management of the market in wine are not in application…producer Member State shall be authorised in order to avoid disturbances on their markets to take measures that may limit imports from another Member State.[53]

The year 1975 was characterized by unusually large flows of Italian wine, particularly into the French market. This was due both to the unusually large grape harvest and to several successive devaluations of the Italian lira. In order to counter what the French government considered to be a disturbance of the French wine market, it imposed an import duty on Italian wine, by reference to Article 31(2) of Regulation (EEC) No 816/70. The duty was imposed by a national decree, in other words a general executive order on the basis of a national law. French importers of Italian wine challenged the validity of the duty before the French courts, from where the case was referred to the Court of Justice for a preliminary ruling.

The Court of Justice referred to Article 38(1) and (2) TFEU, and referred to the fact that the validity of the duty authorized by the Regulation required that legal basis

---

50. On the term 'agricultural products', see Ch. 2, §2.02. The TFEU contains a special rule for State trading monopolies which have rules which are designed to make it easier to dispose of agricultural products or obtain the best return for them.
51. The provisions on the establishment of the internal market for agriculture also involve restrictions on national authorities, individual farmers and the commercial activities associated with agriculture. On the application of the TFEU competition rules to undertakings and the State aid rules, see Ch. 8 below.
52. See Picod, *Agricultural Law for the European Union* 223 (Academy of European Law and Irish Centre for European Law, Heusel & Collins eds., 1999).
53. Regulation (EEC) No 816/70 of the Council laying down additional provisions for the common organization of the market in wine (OJ L 99, 5 May 1970, 1), author's emphasis.

for it could be found in Articles 39–44 TFEU, either expressly or by necessary implication authorizing the introduction of the duty in question. The Court of Justice carefully examined whether these articles contained the necessary provision, but found that 'Articles [39 to 44] contain no provision of this nature'.[54] Article 31(2) of the Regulation was therefore incompatible with the Treaty prohibition of duties on imports and exports and of all charges having equivalent effect, and the Court thus found that it was invalid.[55]

The *Ramel* case furthermore raised the question as to whether Article 38(2) TFEU should be interpreted broadly or narrowly, including the rules on agriculture vis-à-vis the rules on the free movement of goods. The Treaty rules on the common agricultural policy allow for the establishment of support schemes, price-fixing and duties that can affect trade in the internal market. France argued that the free movement of goods was restricted by the mechanisms for the organization of the market, such as price-fixing and intervention systems. France pointed out such limitations were not of a temporary nature nor justified by exceptional circumstances but were characteristic of the common agricultural policy. According to the French argument, the objectives of free movement and of the common agricultural policy should not be set one against the other or put in a hierarchical ranking, but should be combined and the principle of free movement should prevail unless the special requirements of the agricultural sector call for adaptations.[56] The French argument did not appear entirely consistent, but to the extent that it meant that the special Treaty-based rights should be specifically weighed against the common agricultural policy, it was not accepted by the Court of Justice. The Court stated that the prohibition of customs duties on imports and exports and of all charges having equivalent effect was a fundamental principle of the internal market and that it applies to all goods and that 'any possible exception, which in any event must be strictly construed, must be clearly laid down'.[57]

Arrangements for fixed prices, supports and levies have been introduced as part of the common agricultural policy, and especially as parts of common market organizations. These arrangements can be significant barriers to the free movement of goods, but they are expressly authorized by Article 40(2), first subparagraph, TFEU. According to this provision, common organizations of agricultural markets may include all measures required to attain the objectives of the common agricultural policy set out in

---

54. The Court had already established that Art. 39–44 does not contain an exception to Art. 30 in joined Cases 90 and 91/63 *Commission v. Luxembourg* ECR English special edition, 625.
55. Joined Cases 80 and 81/77 *Ramel* [1978] ECR 927, paras 23, 25–27 and 38.
56. Joined Cases 80 and 81/77 *Ramel* [1978] ECR 927, paras 17–19. In the English language version it is assumed that this is the view of the French Government, even if this is not expressed clearly, and it is this that presumably led McMahon, *EU Agricultural Law* 7 and 8–9 (Oxford University Press, 2007), to ascribe this view to the Court. It is clear from e.g., the German language version that the point of view expressed above was in the arguments submitted by the French Government.
57. Joined Cases 80 and 81/77 *Ramel* [1978] ECR 927, para. 24. On the significance of the establishment of a market organization in a given area, see MacMaoláin, *EU Food Law* 56 (Hart, 2007).

Article 39.[58] As possible measures Article 40(2), first subparagraph, names 'in particular regulation of prices, aids for the production and marketing of the various products, storage and carryover arrangements and common machinery for stabilizing imports or exports'.[59]

§3.03 THE ROLE OF THE BASIC PRINCIPLES AND FUNDAMENTAL RIGHTS IN THE INTERPRETATION OF UNION LEGISLATION

The role of the basic principles and fundamental rights in connection with the judicial review of the acts of the institutions is often stated to be a yardstick for the validity of Union legislative acts. Before a case develops to become a question about the validity of a legislative act, the act must be subject to interpretation, and it is here that the basic principles and fundamental rights play a special role as contributors to interpretation or interpretative principles.[60] The basic principles and fundamental rights play a role in interpretation as special factors that influence the understanding of the concrete legislative act. The court has taken the view that when the wording of secondary Union law is open to more than one interpretation, preference should be given to the interpretation which makes the provision consistent with the Treaty rather than an interpretation which leads to the provision being incompatible with the Treaty.[61]

In the *Deuka I* case, the Court of Justice was asked to give a preliminary ruling on whether a Commission Regulation was invalid. Article 1 of the Regulation provided that an arrangement for a higher premium for denaturing wheat would be terminated from 1 June 1970 (the same year), and Article 3 provided that the Regulation entered into force on 15 May 1970.[62] The question was whether the discontinuation of the increased denaturing premium also applied to wheat which had already been purchased by the denaturer before that Regulation came into force. According to the strict wording of the Regulation, it covered wheat that had already been bought but which had not yet been denatured on the date when the Regulation entered into force.

The Court of Justice interpreted the Regulation in the light of the general principle of legal certainty and ruled that the Regulation had to be applied in such a way that

---

58. On the objectives of the common agricultural policy, see Ch. 2, §2.05, above.
59. Compare with McMahon, *EU Agricultural Law* 8 (Oxford University Press, 2007), who point out that common organizations can involve hindrances to trade, but denies that such hindrances can be justified by reference to Art. 36 TFEU.
60. See n. 14 above on the relationship between 'basic principles' and Union legislation. Usher, *EC Agricultural Law* 52–53 (Oxford University Press, 2d ed., 2001), cautiously states that: '[p]erhaps the most frequent use of these general principles in relation to Community agricultural legislation is as a guide to interpretation, so as to ensure the validity of that legislation, rather than as a criterion for determining the validity of that legislation'.
61. Case 218/82 *Commission v. Council* [1983] ECR 4063, para. 15. In Joined Cases 201 and 202/85 *Klensch* [1986] ECR 3477, para. 21, in a case concerning reference quantities in connection with a levy on milk, the Court of Justice stated that when it is necessary to interpret a provision of secondary community law, preference should as far as possible be given to the interpretation which renders the provision consistent with the Art. 34(2) of the EC Treaty (now Art. 40(2) TFEU).
62. Regulation (EEC) No 849/70 of the Commission amending Regulations (EEC) Nos 1403/69 and 1404/69 as regards the denaturing of common wheat (OJ L 102, 12 May 1970, 7).

those who could show that they had bought the wheat in question before the Regulation entered into force were still entitled to the higher premium.[63]

A case can change from concerning a question of the interpretation of Union legislative acts to become a question of the validity of the legislative acts. In the *Hauer* case, the German court's request for a preliminary ruling concerned the interpretation of a Regulation that prohibited the planting of new vines. The Court of Justice ruled, among other things, that the prohibition of new planting also applied to applications for new planting of vines which had been submitted prior to the entry into force of the Regulation.[64] In making its reference, the German court stated that if the Regulation had to be interpreted in this way, then the provision might have to be considered inapplicable in Germany owing to doubts about its compatibility with the fundamental rights guaranteed by German Constitution of the right to property and the right freely to pursue trade and professional activities.[65] This information led the Court of Justice to assess the validity of the prohibition in the Regulation in the light of the same fundamental rights in Union law.[66]

§3.04   THE UNION'S BASIC PRINCIPLES AND FUNDAMENTAL RIGHTS AND NATIONAL AUTHORITIES

The Union's basic principles and fundamental rights not only set limits for the EU institutions. The basic principles and fundamental rights must also be respected by national authorities that implement or apply Union legislative acts in national law. This applies firstly where provisions of Union law are enforced by national courts.[67] Secondly it applies where Treaty provisions allow exceptions to Treaty-based rights.[68] Finally, in a third category of cases, the Union basic principles and fundamental rights

---

63. Case 78/74 *Deuka I* [1975] ECR 421, para. 14. In the French language version it is stated that 'dans l'intérêt de la sécurité juridique, le règlement n° 849/70 devait trouver une application' in the above manner. Hartley, *The Foundations of European Union Law* (Oxford University Press, 7th ed., 2010), states that: '[l]egal certainty – sometimes referred to as "legal security" (*sécurité juridique*, in French) – is another important principle. It is a wide concept which cannot easily be explained in a few words, though predictability is probably the core aspect of it'. See also Case 5/75 *Deuka II* [1975] ECR 759, para. 10, for an interpretation of a regulation in the light of 'legal certainty'.
64. Case 44/79 *Hauer* [1979] ECR 3727, para. 9.
65. *Ibid.* para. 13.
66. See §3.02[E][2] above on the fundamental rights in connection with the question of validity in the *Hauer* case.
67. If, as is often the case, the Union rules leave it to national legislators or administrative authorities to determine the sanctions for infringements of the Union rules, the national sanctions must respect the basic principles and fundamental rights of Union law.
68. In Case C-265/95 *Commission v. France* [1997] ECR I-6995, for a longer period French farmers had committed violent acts and vandalism in order to prevent the import to or transit through France of fruit and vegetables from other Member States, without the French authorities stepping in. The Court of Justice recognized that the Member States had retain exclusive competence as regards the maintenance of public order and the safeguarding of internal security, and that they unquestionably had a margin of discretion in determining the measures most appropriate to eliminate barriers to imports in a given situation. At the same time, the Court stated that, taking due account of the discretion of the Member States, the Court has jurisdiction to verify whether a Member State has adopted appropriate measures for ensuring the free movement of goods; see paras 33 and 35.

must be respected by national authorities which implement Union legislative acts in national law, either by national legislation or national executive orders.

At first sight, the fundamental rights appear to be addressed exclusively to the Union legislator, but they are also binding on national authorities. The *Klensch* case concerned the exercise of discretion left to national authorities in a Regulation establishing a common organization for the market for milk. The Court of Justice referred to the fact that the common market organizations established to implement the common agricultural policy pursuant to Article 40(2), second subparagraph, must exclude any discrimination between producers or consumers within the Union. The Court ruled that this provision applies to all measures related to common market organizations, regardless of the authority which implements them. Consequently, Article 40(2), second subparagraph, is 'binding on the Member States when they are implementing the said common organization of the markets'.[69]

The importance of the obligation of national authorities to respect the fundamental rights of Union law when implementing Union legislative acts in national law increases as the Union rules delegate discretionary powers to national legislators and administrative authorities.[70] For example, the Union legislator can leave it to national authorities to establish agricultural support measures 'on the basis of objective criteria', without stating what these criteria shall be.[71]

The significance of fundamental rights for national authorities was emphasized by the Court of Justice in the *Wachauf* case. Wachauf was a tenant farmer and upon the expiry of his tenancy in accordance with German law he sought compensation from the German authorities for the discontinuance of milk production. The German legislation was based on several regulations that regulated milk production in the European Union. Briefly, the regulations introduced a levy on milk which was payable on quantities of milk from individual farms which exceeded a certain reference quantity. As a rule, the reference quantity for individual farms was set on the basis of the quantity of milk supplied by the farm in 1981. The regulations also provided that a milk producer could seek compensation against an undertaking to discontinue milk production within six months of payment of the compensation. In the cases of tenant farmers,

---

69. Joined Cases 201 and 202/85 *Klensch* [1986] ECR 3477, para. 8. The Court of Justice reinforced this finding by stating that the prohibition of discrimination in Art. 40(2), second subparagraph, is merely a specific enunciation of the general principle of equality which is one of the fundamental principles of Union law; see para. 9. See §3.02[B] above on the principle of equal treatment.

70. McMahon, *EU Agricultural Law* n. 128–130, 29 (Oxford University Press, 2007), emphasizes the importance of the principle of equal treatment where the Union legislator delegates discretionary powers to national authorities.

71. See e.g., Council Regulation (EC) No 1234/2007 establishing a common organization of agricultural markets and on specific provisions for certain agricultural products, Art. 68, Art. 74(1) and (2), and Art. 76(1) and (2); and Council Regulation (EC) No 73/2009 establishing common rules for direct support schemes for farmers under the common agricultural policy and establishing certain support schemes for farmers, Art. 12(3), Art. 28(2), and Art. 41(2), (4) and (6).

they had to submit the landlord's written consent to the discontinuance of production.[72]

The German authorities rejected Wachauf's claim for compensation because the landlord had withdrawn the consent originally given. However, Wachauf maintained his claim for compensation and brought the matter before the German courts, which referred the case to the Court of Justice for a preliminary ruling. In connection with the reference the German court stated that if the regulations were to be understood as meaning that the reference quantity which Wachauf had secured was transferred to the landlord if the landlord refused to give consent, this would mean that Wachauf would be excluded from receiving compensation for the discontinuance of milk production. The German court stated that such a consequence would be unreasonable if, as in this case, the landlord had never produced milk or participated in the establishment of a diary farm. A tenant farmer like Wachauf, who had built up a reference quantity by his own efforts, would be deprived of the fruit of his labours without any compensation. According to the German court, this was contrary to the German Constitution.[73]

As an initial remark in its judgment the Court of Justice stated, with reference to the *Hauer* case, that the fundamental rights form an integral part of the general principles of the law, the observance of which is ensured by the Court.[74] In connection with the actual case, the Court observed that Union rules which, upon the expiry of the lease, had the effect of depriving the tenant of the fruits of his labour and of his investments in the tenanted holding without compensation would be incompatible with the requirements of the protection of the fundamental rights. The Court ruled that: 'Since those requirements are also binding on the Member States when they implement Community rules, the Member States must, as far as possible, apply those rules in accordance with those requirements'. The Court then ruled that the Union regulations in question accordingly leave the competent national authorities a sufficiently wide margin of appreciation to enable them to apply those rules in a manner that is consistent with the protection of fundamental rights, either by giving the tenant the opportunity to keep all or part of the reference quantity, or by compensating him if he undertook to discontinue milk production definitively.[75]

---

72. The relevant rules were in Council Regulation (EEC) No 857/84 adopting general rules for the application of the levy referred to in Art. 5c of Regulation (EEC) No 804/68 in the milk and milk products sector (OJ L 90, 1 Apr. 1984, 13); and Commission Regulation (EEC) No 1371/84 laying down detailed rules for the application of the additional levy referred to in Art. 5c of Regulation (EEC) No 804/68 (OJ L 132, 18 May 1984, 11). On the additional levy scheme in the milk sector see also Case C-45/06 *Campina* [2007] ECR I-2089.

73. Case 5/88 *Wachauf* [1989] ECR 2609, para. 16.

74. See §3.02[E][2] on the *Hauer* case.

75. *Ibid.*, paras 17–23. In Case C-2/92 *Bostock* [1994] ECR I-955, para. 16, on compensation for reference quantities of milk in connection with the termination of a farm tenancy, the Court of Justice repeated that the requirements for the protection of fundamental rights in the Union legal order are also binding on Member States when they implement Union rules so that the Member States must, as far as possible, apply such rules in accordance with those requirements. On the liability of the Member States to pay compensation for failings in the enforcement of the rules on the free movement of agricultural products, see Prieβ in Heusel & Collins (eds.), *Agricultural Law for the European Union* 241 (248–255) (Academy of European Law and Irish Centre for European Law, 1999). On transfer of the reference quantity following the return of part of a holding, see Case C-275/05 *Alois Kibler jun.* [2006] ECR I-10569

If a national authority adopts a legislative act that is contrary to Union law, this can give rise to difficult questions as to whether private individuals can rely on rights arising from such national measures.[76] This is illustrated by the *Deutsche Milchkontor* case which concerned the recovery of aids unduly paid by the German authorities who administered an arrangement under Union law. The aid was paid under an aid measure for the processing of skimmed milk powder, but it subsequently transpired that some of the payments made related to powder that were not made from pure skimmed milk. The Court of Justice emphasized that the principles of the protection of legitimate expectation and assurance of legal certainty are part of the Union legal order. Thus the fact that the national rules protected legitimate expectations and legal certainty in connection with the recovery of unduly paid EU aid could not be regarded as being contrary to the Union legal order. Union law does not prevent national law, in excluding the recovery of unduly paid aid, from having regard to considerations such as the protection of legitimate expectations, or the fact that the administration knew, or was unaware owing to gross negligence on its part, that it was wrong to grant the aid in question. However, this requires that the conditions laid down are the same as for the recovery of purely national financial benefits and that the interests of the Union are taken fully into account.[77]

§3.05 LEGISLATIVE PROCEDURE

Like the other policy areas dealt with in the TFEU, the common agricultural policy has its own provisions governing the legislative procedure to be followed. Article 43(2) and (3) provides that two different legislative procedures are to be used in the area of the common agricultural policy, depending on the matter being legislated on.[78] The starting point is set out in Article 43(2), according to which the ordinary legislative procedure is used to establish the common organization of agricultural markets and the other provisions necessary for the pursuit of the objectives of the common agricultural policy and the common fisheries policy. Article 43(3) provides for a special legislative procedure for the adoption of measures on fixing prices, levies, aid and quantitative limitations. The application of the ordinary legislative procedure in the area of

---

76. On the obligation of national administrative authorities to review a decision (concerning export refunds for poultry meat) in order to take account of an interpretation of Union law given in the meantime by the Court of Justice), see Case C-453/00 *Kühne* [2004] ECR I-837, paras 26–28.
77. Joined Cases 205 to 215/82 *Deutsche Milchkontor* [1983] ECR 2633, paragraphs 30-33. Usher, *EC Agricultural Law* 47 (Oxford University Press, 2d. ed, 2001), is perhaps too categorical when stating that: 'it is clear that the conduct of a national authority which acts in breach of Community law, and has been declared by the Commission to be acting in breach of Community law, cannot give rise to a legitimate expectation'. Compare with McMahon, *EU Agricultural Law* 31 (Oxford University Press, 2007). In the same place, Usher refers to Case C-24/95 *Alcan* [1997] ECR I-1591, para. 25, where the Court of Justice ruled that the recipient of State aid that was contrary to what is now Art. 108(3) TFEU could not have a legitimate expectation about its lawfulness. The *Alcan* case concerned the payment of State aid that was contrary to Union law. The *Deutsche Milchkontor* case concerned the payment of Union aid that was contrary to Union law. On legal certainty in the context of export of cattle, see Case C-2/06 *Willy Kempter* [2008] ECR I-411.
78. This division between the different legislative procedures was introduced by the Lisbon Treaty; see above in Ch. 1, §1.03.

agriculture and the comitology procedure are explained in section §3.05[A], while the special legislative procedure pursuant to Article 43(3) is explained in section §3.05[B].

[A]   The Ordinary Legislative Procedure for the Common Agricultural Policy: Article 43(2) TFEU

The use of the ordinary legislative procedure for the purposes set out in Article 43(2) was an important new element introduced by the Lisbon Treaty. This considerably strengthened the role of the European Parliament in the area of agriculture. Previously the Council merely had to consult the European Parliament. In effect the ordinary legislative procedure corresponds to the legislative procedure which, prior to the Lisbon Treaty, was called the co-decision procedure, as laid down in Article 251 of the EC Treaty. There is a general description of the ordinary legislative procedure in Article 289(1) TFEU, which is in the nature of a definition provision, while the various steps of the procedure itself are set out in Article 294 TFEU. Under this procedure, the Commission has sole competence to submit legislative proposals.[79] The core of the ordinary legislative procedure is that the Council and the European Parliament must agree before legislation can be adopted, meaning that a proposal must receive the support of a majority in both institutions. There can be as many as three phases in the ordinary legislative procedure. A proposal can be adopted in the first phase or it can go through all three phases if there are difficulties in negotiating agreement on a proposal. As soon as a proposal is supported by a majority in both institutions, it is adopted and the procedure is complete.[80]

Pursuant to Article 240(1) TFEU, a committee consisting of the Permanent Representatives of the Governments of the Member States (Coreper) is responsible for preparing the work of the Council. However, at an early stage a rule was made that when the Council deals with questions of agriculture it is the Special Committee on Agriculture (CSA) which prepares the work of the Agriculture Council consisting of the Member States' ministers for agriculture.[81]

[B]   A Special Procedure for Parts of the Common Agricultural Policy, Article 43(3) TFEU

According to Article 43(3) TFEU, the Commission has sole competence to make legislative proposals for measures on fixing prices, levies, aid and quantitative limitations in the area of agriculture. It is the Council alone that adopts such measures. The provision says nothing about whether the Council should consult the European

---

79. The Commission also has sole competence to make proposals in connection with the special legislative procedure referred to in Art. 43(3). This means that the Commission has sole competence to make legislative proposals in the whole of the area of agriculture.
80. On the ordinary legislative procedure, see Hartley, *The Foundations of European Union Law* 36 (Oxford University Press, 7th ed., 2010).
81. The CSA was set up in 1960; see Art. 5(4) of the Decision of 12 May 1960 of the representatives for the governments gathered in Council (OJ 1960, 1217). On the history of the CSA, see Neville-Rolfe, *The Politics of Agriculture in the European Community* 208 (Policy Studies Institute, 1984).

Parliament or how the Council should make its decision. Prior to the Lisbon Treaty, Article 37(2), third subparagraph, of the EC Treaty laid down a legislative procedure under which the Council had to consult the European Parliament before making regulations, issuing directives, taking decisions, or issuing (non-binding) recommendations. Regardless of the kind of legislative act, the Council had to adopt it by a qualified majority.[82] Pursuant to Article 16(3) of the Treaty on European Union (TEU), the Council acts by a qualified majority except where the Treaties provide otherwise. 'Qualified majority' is defined in Article 16(4) and (5) TEU. It must be assumed that the 'measures' referred to in Article 43(3) TFEU is a reference to the kinds of legislative acts referred to in Article 288 TFEU. Within the area covered by Article 43(3) TFEU, the Council is not bound to consult the European Parliament before adopting a measure.

[C] Delegation of Legislative Competence and of Implementing Powers: Comitology

In practice, it is necessary for the Council to delegate legislative competences to the Commission. The common agricultural policy is an area which requires detailed regulation and the frequent issuing of legal measures, among other things to adapt to the changing conditions of the market. In the area of agriculture, prior to the Lisbon Treaty the Council had the practice of issuing framework regulations and directives, authorizing the Commission to adopt implementing legislation. This was owing to that the legislative procedure under the then Article 37(2) of the EC Treaty was far from flexible enough to fulfil the needs for detailed and flexible regulation. According to Article 202 of the EC Treaty, the Council could confer on the Commission powers for the implementation of the rules which the Council laid down, and it could impose certain requirements for the exercise of these powers. Following the Lisbon Treaty, Articles 290 and 291 TFEU distinguish between delegated legislative acts and implementing legislative acts. Article 290(1) provides that a legislative act may delegate to the Commission the power to adopt non-legislative acts of general application to supplement or amend certain non-essential elements of the (delegating) legislative act. By comparison, Article 291(2) among other things provides that, where uniform conditions for implementing legally binding Union acts are needed, those acts shall confer implementing powers on the Commission. According to Article 290(3), the word 'delegated' must be inserted in the title of delegated acts and according to Article 291(4) the word 'implementing' must be inserted in the title of implementing acts.

---

82. Usher, *EC Agricultural Law* 173 (Oxford University Press, 2d ed., 2001), writes that: '[a]lthough … qualified majorities have not always been sought or obtained where they could have been used, they have become the norm in practice'. Originally, the Council made unanimous decisions in the area of agriculture. Article 43(2), second subparagraph, of the EEC Treaty provided that the Council should switch to deciding by a qualified majority in the third stage, i.e., from 1 Jan. 1966; see Art. 8 of the EEC Treaty. However, this transition was prevented by France, whose boycott of the Council resulted in the Luxembourg Compromise. On the common agricultural policy in the light of the Luxembourg Compromise, see Vasey, Com. Mkt. L. Rev. (1988). In 1982 the Council raised agricultural prices by qualified majority over the protests of Denmark, Greece and the United Kingdom; see Hill, *The Common Agricultural Policy: Past, Present and Future* 134–136 (Methuen, 1984).

The distinction between delegated legislative acts and implementing legislative acts is not blindingly obvious.[83] It can be argued that both kinds of legislative acts contain elements of delegation and that both kinds have the character of implementing powers. The further definition of the two kinds of legislative acts must be sought in the case law.

Prior to the Lisbon Treaty, the Council's delegations of legislative competence, broadly understood as covering both delegated and implementing legislative acts, was broadly interpreted in the area of agriculture. Today Article 17(1) TEU provides that the Commission shall ensure the application of the Treaties, and of measures adopted by the institutions, including the Council, to pursue them. Article 17(1) TEU essentially carries forward Article 211 of the EC Treaty, which the Court of Justice interpreted in the *Vreugdenhil* case. The case concerned the exemption from customs duties for a consignment of skimmed milk powder. The Court pointed out that it has consistently held that the concept of 'implementation' in Article 211 of the EC Treaty must be given a wide interpretation. It went on to say that, since only the Commission is in a position to keep track of agricultural market trends and to act quickly when necessary, the Council may confer on it wide powers in that sphere. The Court added that the limits of those powers must be determined by reference to the essential aims of the market organization, and that such a wide interpretation of the Commission's powers can only be accepted in the specific framework of the rules on agricultural markets. However, the Commission cannot claim such a broad interpretation of the provision which it has itself laid down on the basis of its implementing powers in the area of agriculture.[84]

It has not been the intention of the Member States to give the Commission a free rein in conferring powers on it. How the review of the delegated powers is carried out depends on whether a case concerns a delegated legislative act or an implementing legislative act. In the case of a delegated legislative act covered by Article 290 TFEU, the Commission's exercise of its delegation is reviewed by the Council and the European Parliament.[85]

As for implementing legislative acts covered by Article 291 TFEU, a committee system was established and used long before the Lisbon Treaty. This system, usually referred to as the 'comitology procedure' ensures that the Member States can review how the Commission exercises its delegated powers. The delegated powers can only be used by the Commission as long as it has ongoing consultations with a committee that is linked to the exercise of the power in question. The chairmen of these committees are representatives of the Commission, while the other members of the committees are typically civil servants of the national administrations of the Members States. Thus the

---

83. However, in its Proposal for Regulation (EU) No 182/2011, the Commission stated that the Treaty makes a 'clear distinction' (COM(2010) 83 final).
84. Case 22/88 *Vreugdenhil* [1989] ECR 2049, paragraphs 16-17.
85. On the review of the Commission by the Council and the European Parliament, Art. 290 TFEU makes the reservation that these two institutions should only act jointly to the extent that they have acted together in the legislative process. As the ordinary legislative procedure makes these two institutions equal, they exercise their powers jointly under the scope of Art. 43(2). General rules have not yet been adopted for the implementation of Art. 290 TFEU, but see the Commission's thinking in COM(2009) 673 final, and for the European Parliament's view, see document P7 TA(2010) 127.

Member States' governments ensure that they have control over how the Commission uses its delegated powers. The national representatives in each of the committees are instructed by their ministers and report back on how the Commission intends to use its powers in the given case.

The Commission may not delegate the delegated powers that are exercised through the committee system. If this were allowed it would mean that the Commission could get round the system. This is illustrated by the *Rey Soda* case on the regulation of the price of sugar in Italy. Among other things the case concerned the interpretation of Article 6(1) of Commission Regulation (EEC) No 834/74, which provided: 'Italy shall take national measures to prevent disturbances on the market resulting from the increase on 1 July 1974 in the price of sugar expressed in Italian lira. These provisions shall consist in particular of a payment to beet growers of the increased value of stocks.' The Court of Justice ruled that by this provision the Commission had effectively sub-delegated power to the Italian Government. By not specifying in Article 6(1) the bases of the calculation of the tax and by leaving it to Italy to choose them, 'the Commission discharged itself of its own responsibility to adopt the basic rules and to submit them by way of the management committee procedure to the approval if need be of the Council'. Article 6(1) of the Regulation was therefore invalid.[86]

The ordinary rules for the committee system are laid down by the European Parliament and the Council, pursuant to the legal basis in Article 291(3) TFEU, and in Regulation (EU) No 182/2011 laying down the rules and general principles concerning mechanisms for control by Member States of the Commission's exercise of implementing powers.[87] Under Article 2(1) of the Regulation, a binding legislative act, which is referred to in the Regulation as a 'basic act', may provide for the application of either the advisory procedure or the examination procedure. Article 2(2)(ii) provides that the examination procedure is to be used for the adoption of implementing legislation relating to the common agricultural policy.[88] According Article 2(3) of the Regulation,

---

86. Case 23/75 *Rey Soda* [1975] ECR 1279, para. 48–49. The judgment in the *Rey Soda* case is not precedent for that Union legislation cannot delegate powers to the Member States. Union legislation has many examples of such delegations, e.g., in the sugar and milk schemes, and such delegations has been upheld by the Court of Justice in joined cases 103-109/78 *Beauport* [1979] ECR 17. In this case the Council had delegated in a regulation a discretionary power to France to limit or not limit quotas for certain producers of sugar. Compare with McMahon, *EU Agricultural Law* n. 157, 34 (Oxford University Press, 2007), where the *Rey Soda* case is put together with the Beauport case.

87. Regulation (EU) No 182/2011 of the European Parliament and of the Council laying down the rules and general principles concerning mechanisms for control by Member States of the Commission's exercise of implementing powers (OJ L 55, 28 Feb. 2011, 13)

88. The examination procedure also applies to implementing legislation concerning the environment, security and safety, and the protection of the health or safety, of humans, animals or plants. These topics are dealt with in connection with the regulatory procedure; see Art. 5 of Council Decision 1999/468/EC laying down the procedures for the exercise of implementing powers conferred on the Commission (OJ L 184, 17 Jul. 1999, 23). In connection with the amendment, in 2006, of the Council Decision, a special form of 'regulatory procedure with scrutiny' was introduced by Art. 5a; see Council Decision 2006/512/EC (OJ L 200, 22 Jul. 2006, 11). This procedure was to be followed in connection with 'measures of general scope' designed to amend non-essential elements of a legislative act that had been adopted pursuant to the procedure in Art. 251 of the EC Treaty.

as a general rule the advisory procedure applies for the adoption of implementing acts not falling within the scope of application of the examination procedure. However, the advisory procedure may apply for the adoption of the implementing acts where justified.

The advisory procedure corresponds to the previous advisory procedure, while the examination procedure is new, replacing the previous management procedure and regulatory procedure.[89]

As stated above, as a rule the examination procedure applies in connection with the Commission's adoption of implementing legislative acts relating to the common agricultural policy. When exercising the delegated powers, the Commission is assisted by an examination committee composed of representatives of the Member States, with a representative of the Commission as chair.[90] The chair does not have a vote.[91] The Commission representative presents to the committee a draft for the implementing legislation that the Commission proposes to adopt, and the committee then gives its opinion on the draft. If the Member States' representatives cannot agree, the matter is put to the vote. The votes of the Member States' representatives are weighted in accordance with Article 16(4) and (5) TEU, and if appropriate Article 238(3) TFEU. The opinion is thereafter given according to the votes of the majority as laid down in the named Treaty provisions.[92]

If a committee gives a positive opinion the Commission can adopt its proposal for implementing legislation.[93]

If a committee gives a negative opinion, in principle the Commission cannot adopt the legislation.[94] If an implementing legislative act is not regarded as necessary, then pursuant to Article 5(3) the Commission can give up the idea of adopting it.[95] It is not clear from the provision who shall determine whether an implementing legislative act is necessary. In most cases where the Commission has gone to the trouble of preparing draft implementing legislation, one must assume that such an act is necessary. It will usually be more or less clear from the basic legislative act that it is

---

89. See the Commission Proposal for a Regulation of the European Parliament and of the Council laying down the rules and general principles concerning mechanisms for control by Member States of the Commission's exercise of implementing powers (COM(2010) 83 final), 3–4.

90. This will typically be an ad hoc committee convened for the adoption of specific legislation. On specific legal basis, standing committees can take part in an examination procedure pursuant to Regulation (EU) No 182/2011. See for example the Standing Committee on the Food Chain and Animal Health, set up under Art. 58(1) of Regulation (EC) No 178/2002 laying down the general principles and requirements for food law, etc. Art. 48 of Regulation (EU) No 1169/2011 provides that this Committee is a committee within the meaning of Regulation (EU) No 182/2011 and that Art. 5 on the examination procedure applies.

91. Article 3(2) of Regulation (EU) No 182/2011.

92. Ibid., Art. 5(1).

93. Ibid., Art. 5(2). Article 5(4) contains special rules in case the committee does not give an opinion.

94. Article 5(3) refers to Art. 7, which contains a special rule on the Commission's adoption of implementing legislation in exceptional circumstances.

95. In its proposal for Regulation (EU) No 182/2011, the Committee stated that it was a novelty that the Commission did not have an obligation to adopt an implementing legislative act if it was not possible to obtain a qualified majority either for or against a draft measure; see COM(2010) 83 final.

assumed that implementing legislation is necessary. If the Commission fails to put forward a necessary draft, as in a situation where a majority of a committee has given a negative opinion in connection with the examination procedure, depending on the circumstances one or more of the Member States or the Council could bring proceedings against the Commission before the Court of Justice for failure to act. If implementing legislation is necessary, the chair of the committee can either submit an amended draft to the same committee or submit the draft to a special appeal committee.[96] The same procedural rules apply to any amended draft as apply to the first draft, but special procedural rules apply to the submission of a draft to an appeal committee.[97]

For the many cases where a legislative act has been adopted prior to the entry into force of Regulation (EU) No 182/2011, it refers to the rules for the comitology procedure which applied previously, i.e., Council Decision 1999/468/EC. Article 13 of Regulation (EU) No 182/2011 lays down detailed rules for the procedure which should now be followed. In the area of agriculture, in practice the most important cases are situations in which the basic legislative act refers to Article 4 of Council Decision 1999/468/EC. In such cases, according to Article 13(1)(c) of the Regulation, the examination procedure referred to in Article 5 of this Regulation applies.[98]

Article 167(4) of Council Regulation (EC) No 1234/2007 establishing a common organization of agricultural markets and on specific provisions for certain agricultural products, on the granting of export refunds, is an example of legislation on agricultural policy which refers to the former management procedure. According to this provision, the Commission can derogate from certain requirements, but a decision on this must be made using the procedure in Article 16(2) of Council Regulation (EC) No 3448/93, which referred to Article 4 of Council Decision 1999/468/EC on the former management procedure. Now, pursuant to Article 13(1)(c) of Regulation (EU) No 182/2011, the future application of Article 167(4) of Council Regulation (EC) No 1234/2007 should apply the examination procedure.

Experience has shown that it can be very important for individual Member States to seek to persuade the Commission of the correctness of their views before the Commission prepares its draft for a legislative act rather than first giving its opinion during the committee procedure or during the Council's consideration of the draft. In any case, the significance of the rules on amending the Commission's draft legislative acts in connection with the committee procedure should not be exaggerated. In connection with the former Council Decision 1999/468/EC, the Commission emphasized, in a Declaration in connection with the management procedure, that it was its constant practice to try to secure a satisfactory decision which would muster the widest possible support within the committee in question. Moreover, in connection with regulatory procedures the Commission declared that, in order to find a balanced solution in particularly sensitive sectors, the Commission would seek to avoid going

---

96. Article 5(3). The amended version of a draft must be submitted to the same committee within two months of it giving a negative opinion, while the Commission only has one month after a negative opinion has been given to lay the draft before an appeal committee.
97. See Arts 3(7) and 6 on appeal committees.
98. This applies with the exception of Art. 5(4), second and third subparagraphs; see Art. 13(1)(c).

against any predominant position which might have emerged within the Council against the appropriateness of an implementing measure.[99]

Regulation (EU) No 182/2011 contains detailed rules for informing the European Parliament and the Council on the work of the individual committees. To a large extent, these carry forward the rules from Council Decision 1999/468/EC and the agreements between the European Parliament and the Commission associated with it. The Regulation is based on two general principles. First, it is the Member States that have responsibility for reviewing the Commission's exercise of its delegated powers. And second, the procedural requirements should be proportionate to the nature of implementing acts.[100] Article 10(1) of the Regulation requires the Commission to keep a register of committee proceedings, and Article 10(3) states that the European Parliament and the Council must have access to the information in the register. Among other things, the register must record: the agendas of committee meetings; the draft implementing acts on which the committees are asked to deliver an opinion; and the final draft implementing acts following delivery of the opinion of the committees. At the same time as this documentation is sent to the committee members, Article 10(4) requires the Commission to make it available to the European Parliament and the Council, while informing them of the availability of such documents.

Under Article 11 of the Regulation, both the European Parliament and the Council have a right to object to the Commission's draft implementing legislation. Where a basic act is adopted under the ordinary legislative procedure, either the European Parliament or the Council may at any time indicate to the Commission that, in its view, a draft implementing act exceeds the implementing powers provided in the basic act.[101] In such a case, the Commission must review its draft in the light of the positions expressed, and inform the European Parliament and the Council whether it intends to maintain, amend or withdraw the draft implementing act.

---

99. Declarations on Council Decision 1999/468/EC laying down the procedures for the exercise of implementing powers conferred on the Commission (OJ C 203, 17 Jul. 1999, 1).
100. Proposal for a Regulation of the European Parliament and of the Council laying down the rules and general principles concerning mechanisms for control by Member States of the Commission's exercise of implementing powers (COM(2010) 83 final).
101. Under the rules in Council Decision 1999/468/EC, it was only a draft for implementing measures for a legislative act adopted pursuant to Art. 251 of the EC Treaty that had to be sent to the European Parliament. However, the European Parliament and the Commission had a special agreement linked to the Decision, whereby the Commission should also forward drafts for other legislative acts which were of special interest to the European Parliament if the Parliament so requested.

CHAPTER 4
Market Organization

§4.01 ESTABLISHMENT OF A COMMON ORGANIZATION OF AGRICULTURAL MARKETS: ARTICLE 40 TFEU

The EU's agricultural law is based on three pillars, where the market organizations are gathered together as the first pillar.[1] Article 40(1), first sentence, TFEU states that a common organization of agricultural markets is to be established in order to attain the objectives set out in Article 39. While Article 39 sets out in detail the objectives of the common agricultural policy, Article 40 does not contain a definition of a 'common organisation'. In cases on national marketing organizations for bananas, the Court of Justice has defined a national market organization as:

> the totality of legal devices placing the regulation of the market in the products in question under the control of the public authority, with a view to ensuring, by means of an increase in productivity and of optimum utilisation of the factors of production, in particular of manpower, a fair standard of living for producers, the stabilisation of markets, the assurance of supplies and reasonable prices to the consumers.[2]

In all probability, this definition can be extended to apply to market organizations at Union level.[3]

Article 40(1), second subparagraph, states that, depending on the product concerned, a market organization must take the form either of common rules on

---

1. See Ch. 1, §1.01, on the structure and main principles of the EU's agricultural policy and law.
2. Joined Cases 194/85 and 241/85 *Commission v. Greece* [1988] ECR 1037, para. 15; and Case 48/74 *Charmasson* [1974] ECR 1383, para. 22 of the judgment, and para. 2 of the operative part.
3. Barents, *The Agricultural Law of the EC* 44, n. 181 (Kluwer, 1994), states that in the case referred to here the Court of Justice defined 'a common market organisation'. In its Decision 88/109/EEC relating to a proceeding under Art. 85 of the EEC Treaty (IV/31.735 – – New potatoes) (OJ L 59, 4 Mar. 1988, 25), the Commission found that the objectives of the national market organization are similar to those pursued by the common organization at the Community level, as set out in Art. 39 TFEU.

competition, or compulsory coordination of the various national market organizations, or a European market organization. The background to this is that the six original Member States of the European Common Market disagreed about how far agriculture should be subject to a common organization.[4]

Article 40(1) does not require that the whole of the agricultural sector should be regulated in one particular way, but allows the Union legislator to choose according to the individual product type. The provision requires the establishment of a common organization of agricultural markets (plural), not that there should be one market organization (singular) for each type of product.[5] Since, in practice, the Union legislator has chosen to use only the third form (a European market organization), this means that either a product type is subject to a European market organization, or it is at most the subject of certain special Union rules. For example, for many years potatoes, as an agricultural product, were not covered by a European market organization. Potatoes were also not covered by common rules on competition, or compulsory coordination of the various national market organizations pursuant to Article 40(1), second subparagraph, points (a) and (b) TFEU. Thus at Union level potatoes have only been subject to certain special rules. According to Article 182(5) of Council Regulation (EC) No 1234/2007 establishing a common organization of agricultural markets and on specific provisions for certain agricultural products (the 'Single CMO Regulation'), Member States could still provide State aid for potato-growing under existing arrangements until 31 December 2011.[6]

If a product is subject to a common market organization, it can no longer be subject to national regulation. National regulation would require direct authorization in the Treaty or in the common market organization. In proceedings brought for breach of the Treaty, Germany argued that a prohibition in the German food law of the sale of meat products with certain additives, and thus an import prohibition, was necessary for the protection of health and mandatory requirements relating to consumer protection with regard to the common agricultural policy. Among other things, the prohibition affected imports of beef and pig meat that were covered by common market organizations. The Court of Justice ruled that once the Union has established a common market organization in a particular sector, the Member States must refrain from taking any unilateral measure, even if that measure is likely to support the common policy. The Court added that even if they support a common Union policy,

---

4.   Barents, *The Agricultural Law of the EC* 43–44 (Kluwer, 1994), who refers to the *travaux préparatoires* for the EEC Treaty, and is otherwise critical of the possible forms of organization set out in Art. 40(1), second subparagraph, indents (a) and (b). For similar criticism, see McMahon, *EU Agricultural Law* 25–26 (Oxford University Press, 2007).

5.   In Case C-143/91 *Van der Tas* [1992] ECR I-5045, para. 15, the Court of Justice reasoned that Union legislation in the form of directives governing the use of substances having a hormonal effect in livestock farming was 'a common organisation of the market in a particular sector'.

6.   Council Regulation (EC) No 1234/2007. Council Regulation (EC) No 1782/2003 establishing common rules for direct support schemes under the common agricultural policy and establishing certain support schemes for farmers contained certain rules for table potatoes. These rules were carried forward in Council Regulation (EC) No 73/2009. For the use made by individual Member States of the possibility of establishing market organizations for potatoes, see Leidwein, *Europäisches Agrarrecht* 193 (Berliner Wissenschaftsverlag, Österreichischer Agrarverlag and Neuer wissenschaftlicher Verlag Ed. 2, 2004).

national measures may not conflict with one of the fundamental principles of the Union, in this case the free movement of goods, unless they are justified by reasons recognized by Union law.[7]

For many years, common market organizations have been adopted in the form of a regulation for each individual organization. This development has now culminated in the ambition to gather together all market organizations in one regulation.[8] Where a regulation establishing a common market organization includes a directive as part of the organization, the measures which Member States take to give such directives full effect are not characterized as unilateral measures if they are compatible with the aims of the directive they implement.[9]

Article 40(2) TFEU lays down the framework for what a common market organization can include. It is clear that the framework is broad, as it is stated that a market organization 'may include all measures required to attain the objectives set out in Article 39'. As examples, the provision states that 'in particular' such measures may include the regulation of prices, aids for the production and marketing of the various products, storage and carryover arrangements and common machinery for stabilizing imports or exports. The Court of Justice has ruled that the examples given in Article 40(2) are only examples, and that the condition for measures which the Union legislator might wish to introduce is that they should be necessary for achieving the objectives of the common agricultural policy.[10] The term 'all measures' in Article 40(2) covers not only the technical regulation of markets by means of aid, taxes and quotas, but also other kinds of instruments that are relevant for the implementation of the common agricultural policy. Such measures also include the regulation of institutional and formal aspects such as monitoring and the enforcement of union legislation. The Union may decide to restrict itself to requiring the Member States to take the necessary steps to implement Union legislation, or it can choose to lay down more detailed rules for the tasks and powers of national authorities where it finds this more appropriate.[11]

According to Article 288(2) TFEU, regulations are binding in their entirety and are directly applicable in all Member States. As interpreted by the Court of Justice, this means that there is a prohibition on the implementation of regulations by Member States, whereby they are prevented from implementing the provisions of a regulation in national law. This principle does not apply if a regulation specifically requires national implementation of its provisions.[12] This is the case with a number of the provisions in

---

7. Case 274/87 *Commission v. Germany* [1989] ECR 229, paras 21–22. On the significance of Union market organizations for the competences of national authorities, see Usher, European L. Rev. 428 (1977).
8. See §4.02 below on the Single CMO Regulation.
9. Case C-143/91 *Van der Tas* [1992] ECR I-5045, paras. 15–16. Formerly directives were used as the basis for market organizations for individual products.
10. See Case C-240/90 *Germany v. Commission*, [1992] ECR I-5383, para. 18, where the Court of Justice ruled that exclusions from an aid measure as a sanction for economic operators that have committed irregularities in submitting an application for aid was a measure covered by Art. 40(2). See paras 12–13 for references to the Court's case law on the Union legislator's laying down of various forms of sanctions.
11. Barents, *The Agricultural Law of the EC* 47–48 and 209 (Kluwer, 1994).
12. In Case 272/83 *Commission v. Italy* [1985] ECR 1057, para. 27, the Court ruled that where the application in a Member State of Community rules such as those on agricultural producer

the Single CMO Regulation, which provides that national rules are, to varying extents, to fill out the measures for individual products. To a large extent the Member States are required to adopt legislation to enable the national administration of the market organizations.

§4.02   INTRODUCTION TO THE SINGLE CMO REGULATION

[A]   Background and Overview

As stated above in section §4.01, Union legislation previously consisted of a number of basic legislative acts which each established a market organization for a particular product.[13] In 2007, there were 21 market organizations and thus 21 basic legislative acts. The basic legislative acts authorized the Commission to adopt detailed regulations and they generally had a uniform structure. To start with, there was a statement of the extent of the market organization, followed by rules for the internal regulation of the market, and then rules governing the movement of goods between the Union and third countries, and finally there were rules governing any State aid and financing. The instruments used in the various market organizations varied from market organization to market organization, and the regulation was far from being equally intensive in every case. This abundance of regulation has now been pruned and gathered together in one regulation, known as the 'Single CMO Regulation'.[14] The preliminary steps for the Regulation were taken by the Commission in its Communication on Simplification and Better Regulation for the Common Agricultural Policy.[15]

The aim of the Single CMO Regulation was not to implement new political initiatives in the area of agriculture, but to carry out a technical simplification of the regulation of agriculture. Given this, a number of common market organizations that were subject to reform at the time were not included in the original Single CMO Regulation. This originally concerned most parts of the fruit and vegetables, processed

---

groups depends on the combination of a number of provisions adopted at Community, national and regional level, the fact that regional laws incorporate, for the sake of coherence and in order to make them comprehensible to the persons to whom they apply, some elements of the Community regulations, cannot be regarded as a breach of Community law.

13. See Olmi, 'Common Organisation of Agricultural Markets at the Stage of the Single Market', Com. Mkt. L. Rev. 359 (1967–68), for an early description of common market organizations. Prior to the Single CMO Regulation, Grimm, *Agrarrecht*315 (Verlag C. H. Beck, Ed. 2, 2004), stated accurately that '[d]ie Gemeinsamen Marketordnung sind zum Teil sehr kompliziert und unterschiedlich ausgestaltet'.

14. Council Regulation (EC) No 1234/2007 of 22 Oct. 2007 establishing a common organization of agricultural markets and on specific provisions for certain agricultural products (Single CMO Regulation) (OJ L 299, 16 Nov. 2007, 1), which has been amended several times. In French, the Regulation is known as '*règlement OCM unique*', and in German as '*Verordnung über die einheitliche GMO*'. See COM(2006) 822 final, Proposal for a Council Regulation establishing a common organization of agricultural markets and on specific provisions for certain agricultural products.

15. COM(2005) 509 final, 8–9; and the Commission's earlier report to the European Parliament and the Council on the simplification of agricultural legislation (COM(99) 156 final).

fruit and vegetables and the wine sectors. The Single CMO Regulation now includes these sectors or markets.[16]

The Single CMO Regulation not only gathers together the previous common market organizations, it also covers agricultural products that were not previously subject to fully developed market organizations, as well as certain rules which had not been incorporated in market organizations such as the rules on milk quotas, on private storage and on competition and State aid.[17]

The Single CMO Regulation has seven parts. Part I contains provisions on the scope of the Regulation, definitions, and laying down the marketing year for the various sectors of agriculture. Part II is headed 'Internal Market', and contains rules on market interventions and provisions on sales and production. These groups of rules cover: public intervention, private storage, special intervention measures, quotas, aid schemes, marketing standards and conditions for production, as well as rules on producer organizations, inter-branch organizations and operator organizations in the agricultural sector. Part III deals with trade with third countries, including import and export licenses, import duties and levies, inward and outward processing arrangements, and export refunds. Part IV contains competition rules, both rules applying to undertakings and State aid rules. Part V consists of special provisions for individual sectors; and Part VI is headed 'General provisions'. The general provisions include, among other things, authority for the Commission to take the necessary measures in emergencies, rules on the exchange of information between the Member States and the Commission, and a circumvention clause. Part VII, the final part, contains provisions on the use of the committee procedure, transitional provisions and provisions on entry into force.[18]

The entry into force of the Single CMO Regulation means that the common agricultural policy is now governed by four basic regulations adopted by the Council. In addition to the Single CMO Regulation, these are: Council Regulation (EC) No 73/2009 establishing common rules for direct support schemes for farmers; Council Regulation (EC) No 1698/2005 on support for rural development; and Council Regulation (EC) No 1290/2005 on the financing of the common agricultural policy.

---

16. Originally, recitals 7 and 8 and Art. 1(2) of Regulation (EC) No 1234/2007 provided that only Art. 195 of the Regulation (on the Management Committee for the Common Organisation of Agricultural Markets) applied to
the fruit and vegetables, processed fruit and vegetables and the wine sectors. See now recital 3 of Council Regulation (EC) No 491/2009 and recital 8 of Council Regulation (EC) No 361/2008 on the full inclusion in the Single CMO Regulation of the rules for the wine sector and the sectors for the fruit and vegetables, processed fruit and vegetables.
17. COM(2006) 822 final.
18. Article 195(1) and (3) states that the Commission shall be assisted by the Management Committee for the Common Organisation of Agricultural Markets and a legislative committee. Following the Lisbon Treaty, Regulation (EU) No 182/2011 was adopted and references to the management procedure and regulatory procedure must hereafter be read as references to the examination procedure. See Ch. 3, §3.05[C], on the comitology procedure in connection with legislation on agriculture.

[B]   The Scope of Application of the Regulation Etc.

The Single CMO Regulation establishes a single common organization of the markets for the products of a long list of agricultural sectors. The sectors are broadly described in Article 1(1), and specified in more detail in Annex I. Article 1(1), together with Annex I, is important not only because it sets out the scope of the Single CMO Regulation, but also because a number of provisions of the Regulation refer to the Article and to Annex I. The sectors covered by the Regulation are as follows:

- –*cereals*; such as sweetcorn, common wheat, durum wheat, rye, barley, oats, various kinds or flour and various kinds of starch;
- *rice*; including rice in the husk, husked rice, rice flour and rice starch;
- *sugar*; including sugar beet and sugar cane, beet and cane sugar, artificial honey, whether or not mixed with natural honey, and isoglucose;
- *dried fodder*; such as lucerne, clover, lupins, vetches whether artificially heat dried or dried by other means;
- *seeds*; for example peas, beans, spelt or husked rice, all for sowing;
- *hops*;
- *olive oil and table olives*;
- *flax and hemp*; for example flax, raw or processed but not spun, and hemp, raw or processed but not spun;
- *fruit and vegetables*; such as tomatoes, shallots, garlic, carrots, pineapples, table grapes, apples and basil;
- *processed fruit and vegetables*; including vegetables (uncooked, cooked by steaming or boiling in water, or frozen) as well, as dried vegetables, whole, cut, sliced, broken or in powder, dried figs, dried grapes and fruit and nuts, uncooked or cooked;
- *bananas*;
- *wine*; including grape juice (including grape must), wine made from fresh grapes, wine vinegar and wine lees;
- *live plants and products of floriculture*;
- *raw tobacco*;
- *beef and veal*; such as live animals of domestic bovine species, meat of bovine animals whether fresh, chilled or frozen;
- *milk and milk products*; e.g., milk and cream, concentrated and non-concentrated, whey, butter, cheese and cheese curd, and lactose;
- *pig meat*; e.g., live domestic swine, meat of domestic swine, fresh, chilled, or frozen, pig fat, sausages and similar meat products and homogenized preparations of meat, meat offal or blood;
- *sheep meat and goat meat*; including lambs (up to one year old), live sheep and goats, meat of sheep or goats, fresh, chilled or frozen, and fats of sheep or goats;
- *eggs*; poultry eggs, in shell, fresh, preserved or cooked, bird's eggs, not in shell, and egg yolks, fresh, dried, cooked by steaming or by boiling in water, moulded, frozen or otherwise preserved;

- *poultry meat*; live poultry (*gallus domesticus*), ducks, geese, turkeys and guinea fowl, poultry meat and edible offal, fresh, chilled or frozen, and poultry fat;
- *other products*; such as live horses, asses and mules.

Many of the terms used in the Single CMO Regulation, including terms used in Article 1(1), are defined in Article 2 and Annex III of the Regulation. In particular the definitions in Annex III are often highly technical for individual product types and are of major practical importance.[19]

Article 1(3) of the Single CMO Regulation lays down special rules for the market for the agricultural ethyl alcohol sector, the apiculture sector and the silkworm sector, without including these sectors in the common market organization. Similarly, the market for potatoes is subject to special rules; see Article 1(4).

Article 3 lays down six different marketing years for the different product sectors as follows:

Sector	Marketing year
Bananas:	1 January to 31 December the same year
Dried fodder:	1 April to 31 March of the following year
Silkworms:	1 April to 31 March of the following year
Cereals:	1 July to 30 June of the following year
Seeds:	1 July to 30 June of the following year
Olive oil and table olives:	1 July to 30 June of the following year
Flax and hemp:	1 July to 30 June of the following year
Milk and milk products:	1 July to 30 June of the following year[20]
Wine:	1 August to 31 July of the following year
Rice:	1 September to 31 August of the following year
Sugar:	1 October to 30 September of the following year

As for the markets for processed and unprocessed fruit and vegetables, Article 3 gives the Commission powers to determine the marketing year if necessary.[21]

---

19. Article 5(1) and (2) of the Regulation gives the Commission authority to adopt detailed rules for the application of Art. 2, and to amend the definitions of 'rice' and of 'ACP/Indian sugar' in Annex III.
20. In connection with milk quotas, the term 'twelve-month period' is used, beginning on 1 April of the year in question; see §4.04[C][2] below.
21. Article 3 has not yet been used by the Commission.

§4.03   INTRODUCTION TO THE SINGLE CMO REGULATION RULES ON THE INTERNAL MARKET

Part II of the Single CMO Regulation, on the internal market, is the most comprehensive part of the Regulation. It is divided into two titles: Title I, on market intervention, consists of Articles 6–112; Title II, on marketing and production, consists of Articles 113–127. Each title is in turn divided into chapters which are in turn divided into sections and subsections.

Title I of Part II concerns market intervention, which has historically had a leading role in European agricultural policy. Market intervention, also referred to as pricing policy, has traditionally been the central element of common market organizations and thus of the common agricultural policy. The most important market organizations have all been based on the fundamental principle that the prices of the agricultural products covered should be kept at a stable level based on politically determined prices, decided by the Council on a proposal by the Commission. Originally, all prices were set for one year at a time, but following Agenda 2000 and the reforms of several market organizations in 2000 and 2001, the prices of all agricultural products are set for some years ahead.[22] The maintenance of politically determined prices is achieved by the Union's intervention purchases and by the levelling out of any price differences at the Union's outer borders.[23] In recent years, market intervention in the form of pricing policy has been of declining importance for the common agricultural policy. This development is emphasized by the higher priority given to direct aid to individual agricultural producers with limited conditions as to the volume of production and increased rural development aid.

There are four different kinds of market intervention under the current rules. First, there can be public intervention, where the authorities buy up agricultural products or give aid for the private storage of products.[24] Market intervention can also take the form of 'special measures' that only apply in extraordinary circumstances, for example in the event of epidemics of animal diseases or natural disasters.[25] The third form of market intervention can be production limitation. Today, production limitation mainly takes the form of quota schemes.[26] Finally, the Single CMO Regulation contains comprehensive rules on financial support schemes.[27]

Title II of the Single CMO Regulation deals with rules on marketing and production. These rules can be divided into two categories, of which the larger deals with marketing standards for certain kinds of agricultural products, including designations of origin and geographical indications.[28] The second category contains rules on

---

22. On the market organization on common prices and on Agenda 2000, see Ch. 1, §1.01, above; and on the TFEU rules on the price levels of agricultural products, see Ch. 2, in particular §2.05[B][2] and §2.05[B][5].
23. On trade with third countries, see §4.06 below.
24. On this form of intervention, see §4.04[A] below.
25. On this form of intervention, see §4.04[B] below.
26. On this form of intervention, see §4.04[C] below.
27. On this form of intervention, see §4.04[D] below.
28. See §4.05[A] below on the rules for marketing and production.

producer organizations, inter-branch organizations and operator organizations in the agricultural sector.[29]

§4.04 DIFFERENT KINDS OF MARKET INTERVENTION

[A] Market Intervention in the Form of Public Intervention and Storage

The scope of the Single CMO Regulation's rules on market intervention in the form of public intervention and private storage is set out in Article 6(1) of the Regulation. As stated in section §4.03, public intervention consists of the authorities buying agricultural products in order to ensure that the producers are able to sell their goods at a given price level. Private storage consists of giving aid for the private storage of agricultural products. These forms of market intervention can only be used in the sectors of cereals, rice, sugar, olive oil and table olives, beef and veal, milk and milk products, pig meat, and sheep meat and goat meat. Article 7 of the Regulation lays down that only products in the categories named in Article 6 and originating in the Union are eligible for buying-in under public intervention or for the granting of aid for the private storage. Agricultural products in the categories listed in Article 6(1) that are imported from third countries cannot be the subject of the forms of market intervention referred to.[30]

According to Article 7, this rule is stated to apply '[w]ithout prejudice to Article 6(2)'. In Article 6(2), it is emphasized that 'cereals' means cereals harvested in the Union, that 'milk' means cow's milk produced in the Union, and that 'cream' means cream obtained directly and exclusively from milk; see point (d).[31] It is understood that 'cream' exclusively refers to a product originating in the Union, since the cream is derived from 'milk' as so defined. The purpose of Article 6(2) is to define some of the terms used in Article 6(1), and Article 7's reference to Article 6(2) is merely to underline that Article 7's requirements as to statement of origin do not alter the definitions in Article 6(2). This is also clear in language versions other than the English language version, for example the German language version says '[u]nbeschadet des Artikels 6 Absatz 2 kommen nur Erzeugnisse mit Ursprung in der Gemeinschaft', and the French version says '[s]ans préjudice de l'article 6, paragraphe 2, seuls les produits originaires de la Communauté'. Article 6(2) of the Single CMO Regulation repeats word for word Article 5(1) Council Regulation (EC) No 1253/2003 on the common organization of the market of cereals, and Article 6(6)(b) (and the now repealed (c)) of

---

29. See §4.05[B] below on the rules on producer organizations, inter-branch organizations, and operator organizations that are part of the Single CMO Regulation.
30. There was previously a rule that public intervention can only be for the benefit of products originating within the Union in the market organization for beef, see Council Regulation (EC) No 1254/1999 on the common organization of the market in beef and veal, but the fact that the rule has been extended in the Single CMO Regulation means, among other things, that sugar imported under special arrangements for producers in Africa, the West Indies and the Pacific Region, as well as India, cannot be subject to intervention. This is reinforced by Art. 10(1)(c) of the Regulation. See McMahon, *EU Agricultural Law* 383, n. 64 (Oxford University Press, 2007).
31. Article 6(2) originally had a subparagraph (c) with a provision defining 'skimmed milk' as skimmed milk obtained directly and exclusively from cow's milk produced in the Union, but this was repealed by Council Regulation (EC) No 361/2008.

Council Regulation (EC) No 1255/1999 on the common organization of the market in milk and milk products. Point (d) is new, and neither the market organization for cereals nor the market organization for milk and milk products contained as express rules on origins such as those now found in Article 7.

Public intervention and private storage take place on the basis of reference prices. This does not mean that the references prices for individual products are the same as their intervention prices.[32] Article 8 sets out the reference prices for the products listed in Article 6(1). There are special technical specifications linked to some of the prices.[33] Pursuant to Article 8(3), the Council can change the reference prices under the legislative procedure in Article 43(3) TFEU. Changes to the prices therefore require the Commission to make a proposal to that effect and the proposal should receive a qualified majority in the Council.[34]

Article 9 of the Single CMO Regulation contains a special provision on the setting up of an information system on prices in the sugar market.

Article 43 of the Single CMO Regulation contains detailed provisions giving authority to the Commission to adopt a number of provisions for the more detailed implementation of the rules on public intervention and private storage.[35]

[1]   Public Intervention

The products that are eligible for public intervention are listed in Article 10 of the Single CMO Regulation. These are common wheat, durum wheat, barley, maize and sorghum; paddy rice; white or raw sugar; fresh or chilled beef and veal; butter; and skimmed milk powder made by the spray process. Article 10(2) originally contained provision on intervention in the pig meat sector. However, in 2009 it was found that the production and consumption of pig meat could be expected to increase in the medium term, and that pig meat prices could be expected to remain considerably above the intervention price. Since no intervention purchase of pig meat has been made for many years, the Union legislator decided to end the possibility of making intervention purchases of this product.[36] Intervention support for the products covered by Article 10 depends on the fulfilment of a number of conditions which are set out in Articles 10–27 and which are linked to the individual products, as well as any additional conditions which the Commission may lay down.[37]

---

32. On intervention prices see Arts 18-24 of the Single CMO Regulation and immediately below.
33. For example, as regards white sugar and raw sugar, the reference prices are for unpacked sugar, ex factory of standard quality as defined in Art. 8(1)(c) of point B of Annex IV of the Regulation.
34. Article 8(3) of Regulation (EC) No 1234/2007 refers to Art. 37(2) of the EC Treaty, but following the Lisbon Treaty this must be understood as being a reference to Art. 43(3) TFEU. On the legislative procedure under Art. 43(3) TFEU, see Ch. 3, §3.05[B].
35. See Ch. 3, §3.05[C], on the interpretation of the provisions which delegate legislative competence to the Commission
36. Article 10(2) was repealed by Council Regulation (EC) No 72/2009, for the reasons set out in its recital 5.
37. The Commission may lay down such additional requirements and conditions under the authority of Art. 42 of the Single CMO Regulation. See Tiedemann, 'Rechtsprobleme der

There are specific rules for the opening and suspension of intervention purchasing in Articles 11–17 of the Single CMO Regulation. While the Regulation was adopted in 2007, already in 2009 the Union legislator decided to amend significant parts of the rules on public intervention by the adoption of Council Regulation (EC) No 72/2009. These changes are part of the restrictions on the role of pricing policy in the common agricultural policy. Upon its entry into force, public intervention for beef and veal pursuant to Article 14 of the Single CMO Regulation was conditional on the average market price falling below a certain level for a given period. If this occurred, the Commission was bound to open public intervention, while pursuant to Article 17, intervention for pig meat was at the Commission's discretion. For butter and skimmed milk powder, pursuant to Articles 15 and 16 respectively, public intervention was open during the period 1 March to 31 August. However, the Commission could suspend intervention purchases for both products if the quantities offered for intervention exceeded the limits stated in the Regulation. If the Commission suspended public intervention on the basis of the fixed reference prices, purchasing could be continued on the basis of a tendering procedure.[38] As said, in 2009 the Union legislator decided to amend these rules. In respect of cereals, rice, sugar, beef, butter and skimmed milk powder, the current Articles 11–13 provide for both intervention periods and maximum purchase quantities.[39] The rules vary depending on the product in question. There is no maximum quantity for wheat so that according to Article 12(1)(a), during the intervention period for wheat (see Article 11(a)) there can be unlimited public intervention purchases. According to Article 12(1)(b), for durum wheat, barley, maize, paddy rice, sugar, butter and skimmed milk powder, there can only be public intervention purchases up to the maximum quantities laid down in Article 13(1). Under Article 12(1)(c), the Commission can initiate intervention purchasing if the average market price falls below a certain level in a Member State or in a region of a Member State for a representative period.[40]

The intervention prices for the different products are laid down in Article 18. The reference prices for some of the products form the basis or part of the basis for the setting of the intervention price, but as stated previously, the two prices are not necessarily the same. Reference prices play a role in setting the intervention prices for wheat, butter and skimmed milk powder. As part of the restriction of the role of pricing

---

Agrarmarktintervention', RIW 219 (223–228) (1980), on some of the main problems associated with the Union's agricultural rules on public purchases.

38. Other than having different thresholds for their quantities, Art. 15 on butter and Art. 16 on skimmed milk powder had the same content while using different wording. Both provisions were included in the previous Council Regulation (EC) No 1255/1999 on the common organization of the market in milk and milk products. Article 15 had a somewhat different form as Art. 6, while Art. 16(2) of the Single CMO Regulation is clearly based on Art. 7(2). The Art. 15 that was repealed by Regulation (EC) No 72/2009 was inserted in the Single CMO Regulation by Council Regulation (EC) No 361/2008.

39. Originally, the intervention periods for cereals varied between Member States. This was changed by Regulation (EC) No 72/2009.

40. As Art. 13(1)(c) is worded, the Commission is not given any wider discretion, but must initiate purchases when the conditions of the provisions are fulfilled. However, the terms 'in a region of a Member State' and 'representative period' allow the Commission some discretion. See Ch. 3, §3.05[C] on the delegation of powers to the Commission.

policy, an increasing number of the intervention prices for agricultural products are set by tendering. Pursuant to Article 18(2), the intervention prices and intervention quantities for wheat, durum wheat, barley, maize sorghum and paddy rice, beef, veal, butter and skimmed milk powder are set by the Commission by tendering. There is some overlap between the products for which intervention prices are set on the basis of a reference price and the products where the intervention price is set on the basis of tendering. This relates to wheat, butter and skimmed milk powder. The setting of the prices for these products by tendering becomes relevant if the producers, collectively, offer quantities that exceed certain thresholds.[41]

The setting of the intervention price of an individual product by a tendering procedure cannot lead to the price being above a certain level. Under Article 18(3)(a) and (d), the tendering procedure for cereals, paddy rice and skimmed milk powder cannot result in a maximum purchase price that is higher than the reference price for the product in question. As for butter, the maximum purchase price is further limited, as under Article 18(3)(c) it may not be more than 90% of the reference price. The Commission has some discretion in setting the maximum intervention price for beef and veal. Article 18(3)(b) provides that the intervention price may not be higher than the average market price for these products recorded in a Member State or in a region in a Member State, with the addition of an amount which the Commission must set on the basis of objective criteria.

Under Article 18(1), (2) and (3), the intervention price for cereals and paddy rice is subject to the reservation in Article 18(4). For cereals the actual intervention price paid can be higher or lower according to the quality of the product. It is the Commission that decides whether the quality of the cereals can justify an adjustment to the intervention price. In the same way, the Commission can set a higher or lower intervention price depending on the quality of paddy rice offered to the paying agency, as well as ensure that production is orientated towards certain varieties.[42]

Article 18(5) contains a special rule on the intervention price for sugar.

When the Union has made intervention purchases, at some point it must dispose of the products. The Single CMO Regulation lays down three rules on disposals from intervention stocks. Article 25 is headed 'General principles', and it provides that products bought into public intervention must be disposed of in such a way as to avoid any disturbance of the market. Next, disposal must ensure equal access to the goods and equal treatment of purchasers. And finally, disposals must be made in compliance with the Union's obligations under international agreements with third countries or international organizations concluded pursuant to Article 218 TFEU.[43] Article 26 contains special rules on the disposal of sugar, but otherwise the general rules in Article

---

41. For the threshold quantities of wheat, see Art. 18(2)(a), for butter see Art. 13(2)(d), and for skimmed milk powder see Art. 18(2)(e).
42. See Ch. 3, §3.02[A], on the doctrine of the misuse of power.
43. Article 25 of the Single CMO Regulation refers to Art. 300 of the EC Treaty, but following the Lisbon Treaty this has been changed to Art. 218 TFEU, which now determines the procedural rules for all agreements entered into by the Union.

25 apply to all products that can be subject to intervention purchasing. The first principle, that the disposal of intervention stocks must not disturb the market for the product in question was included in several of the former common market organizations.[44] The same was the case with the second principle, on the equal treatment of purchasers, which would apply in any case under the general principle of equal treatment in Union law.[45] The express formulation of the third principle, on compliance with the Union's international obligations, is new.[46]

Article 27 of the Single CMO Regulation contains provisions on the free distribution of products from intervention stocks to the most deprived persons in the Union.[47]

[2]   Private Storage

The rules of the Single CMO Regulation on aid for private storage are in two parts. The Regulation gives authority for the payment of aid for private storage of some products. This is referred to as 'mandatory aid', and it deals with aid for the storage of certain kinds of butter. For other products the aid for storage is 'optional aid', as the aid is dependent on the Commission adopting a decision to pay aid for private storage. The Commission's exercise of this power depends on certain conditions being met. The conditions vary from product to product. For olive oil, pursuant to Article 33 a decision to grant aid for private storage is dependant on the existence a serious disturbance on the market in certain regions of the Union. Under Article 38, the condition of granting aid for private storage for sheep meat and goat meat is that there should be a particularly difficult market situation in one of the areas listed.[48] According to Articles 34–37, for beef products and pig meat, the main condition is that the market price should be less than 103% of the reference price.

It is also an established principle that the paying agencies may only store the products they have bought in the Member State from which they come. This principle can only be departed from if the paying agency in question receives prior permission

---

44. See Art. 6 of Regulation (EEC) No 2759/75 of the Council on the common organization of the market in pig meat; Art. 28 of Council Regulation (EC) No 1254/1999 on the common organization of the market in beef and veal; and Art. 6(4), first paragraph on butter, and Art. 7(4), first paragraph on skimmed milk powder, of Council Regulation (EC) No 1255/1999 on the common organization of the market in milk and milk products.
45. On the general principle of equal treatment in Union law, see Ch. 3, §3.02[B].
46. Article 7(3) of Council Regulation (EC) No 1785/2003 on the common organization of the market in rice contained a rule that under certain circumstances intervention agencies could offer paddy rice for export to third countries; and Art. 7(4), second paragraph, of Council Regulation (EC) No 1255/1999 on the common organization of the market in milk and milk products provided that if skimmed milk powder were to be sold with a view to export, special conditions could be applied to take account of particular requirements in connection with such sale.
47. See §4.04[D][3] below on aid for the supply of milk products to school pupils pursuant to Art. 102, and §4.04[D][7] on Art. 103ga on aid for school fruit schemes.
48. There are also references to particularly difficult market situations in Art. 33 with regard to the storage of olive oil and Art. 45 in the poultry meat and eggs sectors. Article 99 refers to a serious imbalance in the market for skimmed milk and skimmed milk powder and likewise Art. 100 refers to a serious imbalance in the market for casein and caseinates.

from the Commission. Article 39 lays down the more detailed conditions for granting permission for storage in other Member States or third countries.

[B]   Market Intervention: In the Form of Special Intervention Measures

Chapter II of Title I of the Single CMO Regulation, consisting of Articles 44–54, contains the rules on special intervention measures. Where there are restrictions on trade between Member States or with third countries because of measures to combat the spread of diseases in animals, Article 44 gives the Commission authority to adopt exceptional support measures. The Commission may only adopt such measures upon the request of one or more of the Member States affected, and the taking of such measures is conditional on the Member States in question having taken rapid health and veterinary measures to eradicate the disease. Such exceptional support measures should only be taken to the extent and during the period in which it is strictly necessary to support the market in question.[49] This competence of the Commission applies in the beef, milk and milk products, pig meat, sheep and goat meat and poultry sectors.

As for the poultry meat and egg sector, Article 45 gives the Commission authority to adopt exceptional market support measures in order to deal with serious market disturbances directly attributable to a loss in consumer confidence due to public health or animal health risks.

Under Article 47, the Commission may take special intervention measures in the cereals sector where the market situation so dictates. The provision gives the example of situations where, in one or more regions of the Union, market prices fall or threaten to fall in relation to the intervention price. Article 47(2) expressly leaves the nature and application of such measures and the conditions and procedures for the sale or for any other means of disposal of the products to the discretion of the Commission.

In the rice sector, as in the cereals sector, the Commission has wide discretionary powers, as Article 48 states that the Commission may take special measures. The Commission's powers are conditional on the intervention measures being taken to prevent large-scale public intervention in the rice sector in certain regions of the Union, or to make up for paddy rice shortages following natural disasters.

There are detailed rules for special intervention measures in the sugar sector in Articles 49–52. Article 49 lays down minimum prices for sugar beet, to be paid by sugar undertakings that buy sugar beet quotas. The terms for buying sugar beet and sugar cane are governed by agreements pursuant to Article 50(2) within the sector between Union growers of beet and cane and Union sugar undertakings. Agreements within the sector and delivery contracts are determined by the Commission, pursuant to Article 50(1). As an intervention measure that is peculiar to the sugar sector, Article 51 lays down rules for a production charge shall be levied on the sugar quota, the isoglucose quota and the inulin syrup quota held by undertakings producing sugar. Union sugar and inulin syrup undertakings may require sugar beet or sugar cane growers to bear up to 50% of the production charge concerned. The sugar sector also has special rules on the withdrawal of sugar in Article 52 and 52a.

---

49.   On the general proportionality principle of Union law, see Ch. 3, §3.02[C].

In Article 54 of the Regulation, there is a special rule giving the Commission power to take various measures to facilitate the adjustment of supply to market requirements.[50] Such measures can be taken in respect of the live plants, beef and veal, pig meat, sheep meat and goat meat, and eggs and poultry meat sectors. The Commission's powers do not cover measures for withdrawal from the market, but rather:

- measures to improve quality;
- measures to promote better organization of production, processing and market-ing;
- measures to facilitate the recording of market price trends;
- measures to permit the establishment of short and long term forecasts on the basis of the means of production used.

Article 54 gathers together rules which were previously included in the market organizations for the above-named sectors, but what has changed is that such measures are now taken by the Commission pursuant to the Single CMO Regulation, rather than by the Council under the authority of Article 37(2) of the EC Treaty, as previously.[51]

[C] Market Intervention: In the Form of Production Limitation

As the fourth category of the various forms of market intervention, Articles 55–85x of the Single CMO Regulation contain provisions on production limitation. In practice, apart from in the wine sector, production limitation is the same as quota systems. Prior to the adoption of the Single CMO Regulation, there were quota systems in the sugar sector and schemes for reference quantities in the milk sector. The Single CMO Regulation has harmonized the terminology in these two sectors so that the terminology previously used in the sugar sector is now also used in the milk sector, and the terms 'national reference quantity' and 'individual reference quantity' have been replaced by the terms 'national quota' and 'individual quota'. The changes in terminology do not mean a change in substance.[52] In 2009, potato starch was included in the Single CMO Regulation as a product with a quota system.[53] Thereafter, there are quota systems for the following products pursuant to Article 55(1):

---

50. See Ch. 3, §3.02[A], on the doctrine of the misuse of power
51. See the rules previously in Art. 2 of Regulation (EEC) No 234/68 of the Council on the establishment of a common organization of the market in live trees and other plants, bulbs, etc.; Art. 2 of Council Regulation (EC) No 1254/1999 on the common organization of the market in beef and veal; Art. 2 of Regulation (EEC) No 2759/75 of the Council on the common organization of the market in pig meat; Art. 2 of Council Regulation (EC) No 2529/2001 of 19 December 2001 on the common organization of the market in sheepmeat and goatmeat; Art. 2 of Regulation (EEC) No 2771/75 of the Council on the common organization of the market in eggs; and Art. 2(1) of Regulation (EEC) No 2777/75 of the Council of 29 Oct. 1975 on the common organization of the market in poultrymeat.
52. See recital 36 of the Single CMO Regulation.
53. Potato starch was included in the Single CMO Regulation by Art. 4(4) of Council Regulation (EC) No 72/2009. Prior to 2009, the rules on the quota system for potato starch were in Council Regulation (EC) No 1868/94 establishing a quota system in relation to the production of potato

- Milk and milk products.
- Sugar, isoglucose and inulin syrup.
- Potato starch.

Article 55(1)(a) specifies that a quota system shall apply to milk and other milk products as defined in points (a) and (b) of Article 65. Accordingly, 'milk' means the produce of the milking of one or more cows; and 'other milk products' means any milk product other than milk, in particular skimmed milk, cream, butter, yoghurt and cheese.

Article 55(1)(b) does not associate any special definition to the products sugar, isoglucose and inulin syrup, which are covered by the sugar quota system. The same applies to Article 55(1)(c) on potato starch, though this only refers to potato starch for which Union aid can be given.

Regardless of whether it concerns the quota system for milk and other dairy products or the quota system for sugar, isoglucose and inulin syrup, Article 55(2) states that if a producer exceeds the relevant quota, a surplus levy is payable on such excess quantities, subject to the conditions that apply to the individual quota system.

In Article 85, the Union legislator has delegated to the Commission the adoption of the implementing provisions for these three quota systems.

The wine sector is not subject to a quota system but it is regulated by Article 55(2a) and by the provisions in Article 85a–85x on production capacity.[54] Capacity is regulated by rules on unlawful planting, planting rights and a grubbing-up scheme.[55]

[1]   Sugar

According to Article 56 of the Single CMO Regulation, national and regional quotas for the production of sugar, isoglucose and inulin syrup are at the levels laid down in Annex VI. For example, for marketing year 2010/2011, the United Kingdom had a quota of 1,056,474 tons of sugar, but no quota for isoglucose; Germany had a quota of 2,898,255.7 tons of sugar, and 56,538.2 tons of isoglucose; France had a quota of 3,004,811.15 tons of sugar. At the other end of the scale, a smaller country such as Denmark had a quota of 372,383 tons of sugar and no quota for isoglucose. No Member State has a fixed quota for inulin syrup. Under its authority in Article 59, and subject to certain conditions, the Commission can adjust the quotas in Annex VI.

The individual Member States allocate the national (or regional) quotas to the undertakings producing sugar, isoglucose or inulin syrup that are located in their territories. Only undertakings which the individual Member State has approved pursuant to the conditions in Article 57 of the Regulation can be allocated a quota.

---

starch. See recital 15 of Regulation (EC) No 72/2009 for the background to the repeal of Regulation (EC) No 1868/94 and the inclusion of potato starch in the quota systems of the Single CMO Regulation.

54. These provisions were incorporated into the Single CMO Regulation by Council Regulation (EC) No 491/2009, in connection with the full integration of the wine sector under the Single CMO Regulation.

55. Quotas for new planting rights are laid down for each Member State; see §4.04[C][4] below.

According to Article 57(1), approval requires an undertaking to prove its professional production capacities and to agree to provide any information and to be subject to controls related to the Regulation. Under Article 60(1), a Member State may reduce the sugar or isoglucose quota it has allocated to an undertaking by up to 10% for each marketing year. In Article 60(2), there is authority for Member States to transfer quotas between undertakings, taking into consideration the interests of each of the parties concerned, particularly the interests of sugar beet and cane growers. Annex VII gives detailed rules on Member States' transfers of quotas between undertakings.[56]

As stated in section §4.04[C], producers who exceed their quotas have to pay a surplus levy. However, in the sugar sector over-production does not automatically trigger a levy as Article 61(a) to (d) gives individual undertakings several options for dealing with the excess so that a levy need not be paid. One option is for an undertaking to process the excess production into one of the products referred to in Article 62, such as bioethanol and alcohol. Another option is for an undertaking to carry forward its excess production to be treated as part of the next marketing year's production, as set out in Article 63. There are certain time limits for such a decision, and it is a condition that the undertaking stores the surplus production at its own expense until the end of the current marketing year. A third possibility is for the undertaking to enter into a specific supply regime for the EU's outermost regions. Finally, the undertaking can export that part of its production that exceeds its quota to third countries. The use of this option presupposes that such export can take place within the quantitative limit which the Commission has established pursuant to the Union's international obligations.[57]

Quantities of sugar that are produced in excess of the allocated quota and which are not used in one of the ways referred to in Article 61(a) to (d) will trigger the obligation to pay a surplus levy pursuant to Article 64.[58]

Article 64(2) states that it is the Commission that determines the amount of the surplus levy, and that it should be at a sufficiently high level to prevent the accumulation of surpluses. Article 64(3) requires the Member States to collect the surplus levies from the undertakings.

[2] Milk and Other Milk Products

The national quotas for the production of milk and other milk products are laid down in Annex IX of the Single CMO Regulation. For example, for the 12-month period during 2012/13, the United Kingdom has a quota of 15,739,311.451 tons; Germany has a quota of 30,018,741.337 tons; France has a quota of 26,110,129.977 tons; while a small Member State like Denmark has a quota of 4,799,910.369 tons. The quotas apply for a

---

56. Cf. Art. 74 on the transfer of milk quotas (see §4.04[C][2]) and Art. 85i on the transfer of planting rights for vines (see §4.04[C][4]).
57. Article 61(d) refers to the commitments of the Commission 'resulting from agreements concluded in accordance with Article 300 of the Treaty'; following the Lisbon Treaty this must be read as a reference to Art. 218 TFEU.
58. On the calculation of production levies in the sugar sector, see joined Cases C-5/06 and C-23-36/06 *Zuckerfabrik Jülich AG and others* [2008] ECR I-3231.

12-month period, beginning on 1 April. It is the intention of the Union legislator that milk quotas should come to an end in 2015.[59]

As with sugar quotas, national milk quotas are allocated by the Member States to individual producers in their territories. The introduction of milk quotas instead of reference quantities by the Single CMO Regulation did not mean that the milk quotas could or should be reallocated. Article 67(1) provides that individual producers' quotas at 1 April 2008 should be the same as their individual reference quantities at 31 March 2008 under the previous market organization for milk and milk products.[60]

Under Article 66(2), individual quotas allocated to producers can either be for deliveries or direct sales. Article 67(2) provides that producers may have either one or two individual quotas, one for deliveries and the other for direct sales. In Article 65(f), 'delivery' is defined as any delivery of milk, not including any other milk products, by a producer to a purchaser, regardless of who is responsible for the transport. In Article 65(g) 'direct sale' is defined as any sale or transfer of milk by a producer directly to consumers, as well as any sale or transfer of other milk products by a producer. Among other things, the use of the term 'direct sale' is intended to ensure that producers cannot circumvent the quota system by selling milk and other milk products directly to consumers, by-passing 'purchasers' and thus outside the national milk quota. For this purpose, Article 65(g), second sentence, authorizes the Commission, while respecting the definition of 'delivery' in point (f), to adjust the definition of 'direct sale' in order to ensure that no quantity of marketed milk or other milk products is excluded from the quota arrangements. Article 69 requires the Commission to adjust the division between 'deliveries' and 'direct sales' in national quotas for each Member State and for each period.

Under the provisions in Article 78, if milk and other milk products are marketed in a Member State in excess of the national quota, the Member State must pay a surplus levy to the Union.[61] In Article 78(1), the levy was set at EUR 27.83 per 100 kilograms of milk. Amendments to the Single CMO Regulation raised this considerably for the 12-month periods beginning 1 April 2009 and 1 April 2010, but not for subsequent 12-month periods.[62] Article 79 requires the Member States to allocate the surplus levy among the producers who have contributed to each of the overruns of the national quotas and these producers must pay the Member State for their share of the surplus levy. In calculating any surplus levy, a distinction is made between exceeding a quota for deliveries and exceeding a quota for direct sales. If a producer has two quotas, under Article 67(3) the contribution to any surplus levy due is to be calculated separately for each. Articles 80–82 contain relatively technical rules on the calculation of individual producers' share of the total surplus levy in respect of surplus levy for a

---

59. See e.g., recital 8 in Council Regulation (EC) No 72/2009.
60. See above §4.02[C] on the introduction of milk quotas instead of reference quantities.
61. Surplus levies raised pursuant to Art. 78 are regarded as 'assigned revenue', as referred to in Art. 34(1)(b) of Council Regulation (EC) No 1290/2005 on the financing of the common agricultural policy.
62. On the background to the increases in the surplus levy for the two 12-month periods referred to, see recital 8 in Council Regulation (EC) No 72/2009 and recital 2 in Council Regulation (EC) No 1140/2009.

producer's deliveries based on the reference fat content (see Article 70), and Article 83 in respect of surplus levy for direct sales.

'Purchasers' play a special role in calculating and collecting producers' shares of any surplus levy in connection with the delivery of milk and other milk products. A 'purchaser' is defined in Article 65(e) as an undertaking or group which buys milk from producers, and it is the purchaser which, according to Article 81, has responsibility for collecting the contributions which producers must make to a surplus levy. The purchaser pays the levy collected to the competent body of the Member State. Article 82 states that the status of 'purchaser' is subject to prior approval by the Member State in question in accordance with criteria laid down by the Commission.[63]

Even though Article 67(1) refers to 'producers' individual quota or quotas' and Article 67(2) states that '[p]roducers may have either one or two individual quotas', the quota is in fact linked to the farm, and if a producer transfers a farm to another producer Article 74(1) states that individual quotas are to be transferred with the holding to the producers taking it over when it is sold, leased, transferred by actual or anticipated inheritance or any other means having comparable legal effects for the producers. It is up to the Member States to lay down more detailed regulations for the transfer of quotas in such circumstances.

Where quotas are transferred by means of rural leases or by other means involving comparable legal effects, Member States may lay down rules whereby the quota is not transferred with the holding. Article 74(2) makes such rules conditional on such criteria having the aim of ensuring that quotas are allocated solely to producers.[64]

A number of factors can mean that the basis for the allocation of a quota no longer exists. As the individual quotas collectively constitute the national quota, by gathering together such dormant milk quotas a Member State can establish a national reserve. Under Article 71, individual Member States may elect to supplement their national reserve by withdrawing quota allocations referred to in Article 72 on the cessation of activities by withholding part of the quota transferred between producers in connection with the transfer of a holding, or by making an across-the-board reduction of all individual quotas. Article 68 provides that each Member State may adopt rules for the allocation of quotas from the national reserve to producers. According to Article 77, Member States may not provide financial assistance directly linked to the acquisition of quotas by sale, transfer or allocation.[65]

If a producer ceases to produce and sell milk, and this cessation lasts for more than a 12-month period, Article 72(1) provides that the producer's quota shall revert to the national reserve. Where a producer does not market a quantity equal to at least 85% of their individual quota during at least one twelve-month period, Member

---

63. For the conditions which the Commission has laid down for approval of purchasers, see Art. 23 of Commission Regulation (EC) No 595/2004 laying down detailed rules for applying Council Regulation (EC) No 1788/2003 establishing a levy in the milk and milk products sector.

64. Cf. Art. 60(2) on the transfer of sugar quotas, in §4.04[C][1] above, and Art. 85i on the transfer of replanting rights for vines, below in §4.04[C][4]. On temporary acquisition of a reference quantity by a lessor who is not, and does not intend to become, a milk producer, see Case C-278/06 *Otten* [2007] ECR I-4513.

65. On State aid to undertakings in the agricultural sector, see Ch. 8, §8.04.

States may decide whether and on what conditions all or part of the unused quota must revert to the national reserve.[66] In these cases, the special rules on the reversion of quotas from the national reserve to the producer in question apply.[67]

[3]   Potato Starch

The quota system for potato starch is very simple compared with the arrangements for sugar and milk. Article 84a(1) of the Single CMO Regulation, as amended by Council Regulation (EC) No 72/2009, provides that the Member States listed in Annex Xa shall be allocated quotas for the marketing of potato starch.[68] Under Article 84a(2), each producer Member State allocates its quota among potato starch manufacturers for use in the marketing years concerned. These allocations are made on the basis of the sub-quotas allocated to each manufacturer in 2007/2008.[69] Article 84a(3) provides that potato starch producers may not enter into cultivation contracts with potato producers for a quantity of potatoes which would produce a quantity of starch in excess of their quotas. If a potato starch producer nevertheless produces more than its allocated quota, under Article 84a(5) it may use no more than 5% of its quota relating to the following marketing year, and the following year's quota is reduced accordingly. If a potato starch producer chooses not to carry forward its excess production into its quota for the following year, or if it exceeds its quota by more than 5%, the potato starch produced in excess of the quota must be exported from the Union before 1 January following the end of the marketing year in question. The provision states clearly that in this case no export refund may be paid in respect of it.

According to Article 84a(6), potato starch producers who have not been allocated a quota pursuant to Article 84a(2), and who buy potatoes from potato growers who do not receive aid pursuant to Article 77 of Regulation (EC) No 73/2009, are not subject to the Single CMO Regulation's potato starch quota system.

[4]   Wine

As stated in section §4.04[C], the wine sector is regulated by the Single CMO Regulation in the form of production limits. In contrast to other sectors that are also covered by the Regulation, there is no quota system for the production of wine. According to Article

---

66.   Originally, the Single CMO Regulation set this percentage at 70%, but this was altered by Council Regulation (EC) No 72/2009. The reason for this change was the desire to make the milk quota system more flexible; see recital 8 of Regulation (EC) No 72/2009.
67.   On the case law on milk quotas prior to the Single CMO Regulation, see Cardwell, 'General principles of community law and milk quotas' Com. Mkt. L. Rev. 723 (1992), and Cardwell, *The European Model of Agriculture* 252–256 (Oxford University Press, 2004); and O'Reilly, *Agricultural Law for the European Union* 101 (Heusel & Collins eds., Academy of European Law and Irish Centre for European Law, 1999).
68.   Article 84a(1) contains a reference to Art. 204(5) of the Single CMO Regulation with a provision that potato starch quotas only apply until the end of marketing year 2011/12. For the reasons for the time limit for the quota system, see recital 15 of Regulation (EC) No 72/2009.
69.   Among other things, compliance with the quota determines whether aid is paid for processing pursuant to Art. 95a; see §4.04[D][1] below.

55(2a) of the Single CMO Regulation, as amended by Council Regulation (EC) No 491/2009, the Regulation seeks to regulate the production of wine through three kinds of measures. The first concerns the rules on unlawful planting in Article 85a–85e. The second concerns rules on transitional planting rights in Article 85f–85n.[70] Finally, there are rules for a grubbing-up scheme in Article 85o–85x.

The production rules for wine in the Single CMO Regulation are based on the same rules as were found in Council Regulation (EC) No 479/2008 on the (previous) common organization of the market in wine.[71] Both the Single CMO Regulation and Regulation (EC) No 479/2008 are silent on the background to the rules, and these rules may reasonably be characterized as for insiders. The rules refer to 'planting rights' without giving any explanation of what is meant by this, but according to Article 85g(3), regardless of whether there are rules on unlawful planting or transitional planting rights, there can be new planting rights, replanting rights, and planting rights granted from a reserve.

The aim of the Single CMO Regulation and the previous legislative acts regulating planting in the wine sector has been to control the areas under cultivation. The rules date back to Council Regulation (EEC) No 1162/76 on measures designed to adjust wine-growing potential to market requirements, which was adopted in order to counter 'a considerable imbalance in the table wine market'.[72] Article 2(1) introduced a prohibition of all planting of new vines.[73] The prohibition was for a limited period, but it was subsequently extended and the prohibition is now expressed in Article 85g of the Single CMO Regulation. The planting prohibition in Article 85f extends until 31 December 2015.[74]

Like the previous rules, the existing rules are based on the principle that the planting of vines is prohibited without the existence of a replanting right or a new planting right. Broadly speaking, these rules have been in use in the EU since the end of the 1970s. However, Council Regulation (EC) No 1493/1999 on the common organization of the market in wine made changes to the earlier market organization which hang together with the economic development of the sector. The two most important changes were the setting of quotas for new planting rights for each Member State, and the establishment of reserves of planting rights.[75]

As stated above, Article 85g imposes a transitional prohibition on the planting of vines. The prohibition also covers the grafting-on of wine grape varieties to non-wine

---

70. According to Art. 85l, Arts 85f to 85n (i.e., subsection II) only apply in those Member States where the Unions planting rights scheme applied as at 31 Nov. 2007.
71. As stated in footnote 54, the rules regulating production in Regulation (EC) No 479/2008 were fully incorporated in the Single CMO Regulation by Regulation (EC) No 491/2009.
72. See the first recital of Regulation (EEC) No 1162/76.
73. There was a single exception to the prohibition in Art. 2(2), but this did not alter the fact that the prohibition was – and is – the most direct form of production regulation. See Snyder, *Law of the Common Agricultural Policy* 133 (Sweet & Maxwell, 1985). The validity of the Regulation was challenged in Case 44/79 *Hauer* [1979] ECR 3727, but without success.
74. Article 85g(5) allows individual Member States to retain prohibitions of planting and grafting until 31 Dec. 2018.
75. Commission report to the European Parliament and the Council on management of planting rights (COM(2004) 161 final, 3).

grape varieties. The prohibition of planting and grafting only applies to grape varieties classified under Article 120a. Under Article 120a(2), it is the Member States that classify which wine grape varieties may be planted, replanted or grafted on their territories for the purpose of wine production. The Member States do not have unfettered discretion to classify, as the provision both makes positive requirements as to the varieties of grapes and negative requirements, excluding certain varieties.[76]

According to Article 85g(3), the prohibition of planting and grafting does not apply where there is:

- a new planting right;
- a replanting right;
- a planting right granted from a reserve.

The rules for the allocation of new planting rights are in Article 85h. The rights are allocated to individual Member States, but the general prohibition of planting in Article 85g means that new planting rights can only be granted within narrow limits. Thus, rights can be granted in connection with measures concerning compulsory purchases, where the planting is for experimental purposes, for graft nurseries, or where the wine or vine products are intended solely for consumption by the wine-grower's household. The fact that even new planting for private use requires a Member State's authorities to grant the producer a new planting right underlines the wide scope of the planting prohibition under Article 85g. New planting rights cannot be assigned, as Article 85h(2) states that the rights must be exercised by the producer to whom they are granted and must be used for the purposes for which they are granted.

According to Article 85i(1), Member States may grant replanting rights to producers who have grubbed up an area planted with vines and, according to Article 85i(2), also to producers who undertake to grub-up an area planted with vines.[77] Article 85i(4) provides that replanting rights must in principle be exercised on the holding in respect of which they are granted. The provision also states that Member States may stipulate that such replanting rights may be exercised only on the area where the grubbing-up was carried out. As an exception to the main rule, replanting rights can in some cases be transferred to another holding.[78] However, under Article 85i(5), it is a general condition that the Member State in question should have established rules for this and that the transfer is made to another holding in the same Member State. Member States can only permit the transfer of rights as long as it does not lead to an increase of the production capacity in their territory. Under these circumstances, replanting rights can be transferred if part of the holding concerned is transferred to that other holding. Transfer is also possible if the areas on that other

---

76. Under Art. 120a(3), Member States whose wine production does not exceed 50,000 hectolitres per wine year, calculated on the basis of the average production during the latest five wine years, are exempt from the classification obligation. However, these Member States are subject to other restrictions on the varieties of grapes that may be cultivated.
77. Planting rights pursuant to Art. 85i(2) involve the 'co-existence of vines'; see Art. 85n(1)(b).
78. Compare with Art. 60(2) on the transfer of sugar quotas, above in §4.04[C][1], and Art. 74 on the transfer of milk quotas, above on §4.04[C][2].

holding are intended for the production of wines with a protected designation of origin or a protected geographical indication, or for the cultivation of graft nurseries.

Even though there has been a planting prohibition under Union law for many years, and the Member States have been required to ensure compliance with the prohibition, the reality is that over the years there has been a lot of unlawful planting of vines. The Union legislator seeks to eradicate unlawful planting by the measures in Article 85a–86e. Article 85a deals with unlawful plantings planted after 31 August 1998, and Article 85b deals with unlawful plantings planted before 1 September 1998.

Article 85a(1) provides that producers who have planted areas with vines after 31 August 1998, without a corresponding planting right, must grub-up such areas at their own expense. Under Article 85a(3), Member States are required to impose penalties on producers who have not complied with this grubbing-up obligation; such penalties must be graduated according to the severity, extent and duration of the non-compliance. The Union legislator has presumably not had full confidence in unlawful plantings being grubbed up immediately, as Article 85a(2) provides that grapes and products made from grapes from unlawfully planted areas may only be put into circulation for the purposes of distillation. Such distillation must be at the exclusive expense of the producer, and products resulting from distillation may not be used in the preparation of alcohol having an actual alcoholic strength of 80% by volume or less.

As for unlawful planting dating from prior to 1 September 1998, Article 85b contains rules for their obligatory regularization. Not later than 31 December 2009, producers had to regularize areas planted with vines without a corresponding planting right prior to 1 September 1998 by the payment of a fee. The fee was to be determined by Member States, but Article 85b(2) required the fee to be at least equivalent to twice the average value of the corresponding planting right in the region concerned. Areas that had not been regularized by 31 December 2009 had to be grubbed up by the producer concerned at their own expense. Until the grubbing-up required by the law was carried out, Article 85b(3) imposed a distillation rule corresponding to that in Article 85a(2).

Where there is production on the basis of unlawful planting, pursuant to Article 85c(1) the Member States have to require proof of non-circulation of the products concerned or, where the products concerned were distilled, the submission of distillation contracts. In extension of this requirement for documentation, Article 85c(2) requires the Member States to verify such non-circulation or distillation. On the basis of the independent monitoring obligation imposed on the Member States by Article 85c(2), it must be assumed that the Member States cannot merely be satisfied by receiving documentation of compliance with the prohibition of circulation or requirement for distillation, but must carry out some other form of control. Under Article 85c(3), the Member States must notify the Commission of the areas subject to distillation and the corresponding volumes of alcohol.

In Article 85e, the Union legislator has given the Member States power to lay down implementing measures, including provisions on the penalties which the Member State will impose on producers for failure to comply with the rules on unlawful planting, on the prohibition of putting products from unlawfully planted areas into circulation, and on the requirement to distil such products.[79]

As with quota schemes for other agricultural products, it was found appropriate to lay down rules for reserves of planting rights in connection with the planting of vines. Article 85j states that, in order to improve management of the production potential, Member States should create national reserves or regional reserves of planting rights.[80] As a special exception, Article 85j(5) provides that Member States may decide not to implement a reserve system provided that they can prove that an effective alternative system for managing planting rights exists throughout their territory. While the exception provision refers to Member States in the plural, it must be understood as meaning that it is up to the individual Member State to decide whether some arrangement other than a reserve system should be used. If a Member State does choose some other system, it can derogate from the rules that apply to reserve rights.

According to Article 85j(3), a reserve of planting rights, whether national or regional, gathers together planting rights that have not been used within the prescribed period. This applies to new planting rights, replanting rights, and planting rights previously granted from the reserve.[81] Moreover, according to Article 85j(4), producers may transfer replanting rights to national or regional reserves in return for a payment from the Member State. It is up to the Member State to determine the conditions for such a transfer.

All planting rights in a reserve can be withdrawn from the reserve in one of three ways. First, pursuant to Article 85k(1)(a), Member States may grant rights from a reserve, without payment, to producers who are under 40 years of age, who possess adequate occupational skills and competences, who are setting up for the first time and who are established as the head of the holding. Second, pursuant to Article 85k(1)(b), planting rights can be granted from the reserve against payment to national or regional funds, to producers who intend to use the rights to plant vineyards which have an assured outlet for their production. In this case, the Member State concerned defines the criteria for setting the amounts of the payment, which may vary depending on the final intended product of the vineyards concerned and on the residual transitional period during which the prohibition of new planting applies.[82] Finally, planting rights can be removed from the reserve if they are not used in due time. Article 85k(4) states that planting rights in a reserve which are not disbursed before the end of the fifth wine year following their allocation to the reserve are extinguished.

---

79. Commission Regulation (EC) No 555/2008 laying down detailed rules for implementing Council Regulation (EC) No 479/2008 on the common organization of the market in wine etc.
80. Where a Member State decides to establish several regional reserves, it can allow transfers between regional reserves and between regional reserves and the national reserve; see Art. 85k(5).
81. Under Art. 85k(3), planting rights granted from a reserve which are not used before the end of the second wine year after the one in which they were granted are forfeited and reallocated to the reserve.
82. On the interim prohibition of new planting of vines, see Art. 85g(1) and (2), and footnote 74 above.

In the first two cases, where planting rights from reserves of planting rights are used, Article 85k(2) requires the Member States to ensure that such grants of rights do not lead to over-production. On the one hand, the Member States must ensure that the location, varieties and cultivation techniques used guarantee that the subsequent production is adapted to market demand; and on the other hand they must ensure that the yields concerned are typical of the average in the region. This latter applies in particular where planting rights originating in non-irrigated areas are used in irrigated areas.

Article 85m gives the Member States authority to adopt stricter national rules in respect of the award of new planting rights or replanting rights. Member States may require applications and the information supplied in them to be supplemented by additional information necessary for monitoring the development of production potential in the State. The scope of the provisions are not expressly limited to the rules in Article 85f–85n, i.e., subsection II, but it must be assumed that authority to impose stricter national rules only applies to these provisions.

Article 85n gives the Commission authority to adopt detailed rules for the implementation of the transitional planting rights regime, i.e., Article 85f–85n.[83]

As stated in the introduction to this section, the regulation of wine production is exercised by three arrangements, of which the grubbing-up scheme in Article 85o–85x is the third.[84] The grubbing-up scheme pursuant to Article 85o applied until the end of the wine year 2010/2011. It was a scheme under which a premium was paid to vine growers for the grubbing-up of vines, referred to in Article 85p as a 'grubbing-up premium'. According to Article 85q, there were a number of conditions attached to the payment of the premium, and under Article 85r the scales for the grubbing-up premiums were fixed by the Commission, while the specific amount of the grubbing-up premium was established by the Member State in question.[85]

The grubbing-up scheme was characterized by the existence of a cross compliance requirement.[86] Under Article 85t, it was a condition for the payment of a grubbing-up premium that at any time during three years from payment of the grubbing-up premium the recipient complied on the holding with the statutory management requirements and the good agricultural and environmental condition referred to in Articles 3–7 of Regulation (EC) No 1782/2003.

[D] Market Intervention: Aid Schemes

The Single CMO Regulation Title on market intervention contains a chapter on aid schemes, governing 10 such different schemes. Aid takes different forms in different sectors. This aid should not be confused with the support that is paid in connection with trade with third countries.

---

83. The Commission exercised this authority in Commission Regulation (EC) No 555/2008.
84. See Art. 55(a) of the Single CMO Regulation. In Art. 120a(1) and (5) there are rules on the grubbing-up of vines in connection with declassifying a grape variety; see §4.05[A][12] below.
85. For the scales determined by the Commission, see Commission Regulation (EC) No 555/2008, Annex XV.
86. On cross compliance in general, see Ch. 7.

[1]   Aid for Processing

Aid is paid for the processing of two groups of agricultural products. First, aid is paid for the processing of flax and hemp for the production of fibre. Second, aid is paid, pursuant to Article 95a, for the production of potato starch. As will be clear from the following, processing aid is a form of aid that is still in the development stage.

Until 1 April 2012, there was yet another scheme for processing aid under the Single CMO Regulation. Articles 86–90 laid down rules for aid for drying fodder.[87] In 2009, in Council Regulation (EC) No 72/2009, the Union legislator decided to do away with this scheme as part of the decoupling of agricultural production support. As reasons for this decision, the Union legislator referred to the overall move in Union agricultural law towards more market orientation, the outlook in the markets for feed and protein crops, and the particular negative environmental impact that the production of dehydrated fodder had been found to generate. The legislator has sought to soften the impact of the removal of this aid for farmers by adjusting the price paid to the producers of raw materials who have been given increased rights to direct aid as a consequence of the decoupling. Regulation (EC) No 72/2009 entered into force on 1 April 2012, so that the sector could have the possibility of making adjustments.[88] Article 86 laid down requirements for eligibility for both the undertaking and the fodder.[89] Article 88 established the level of aid at EUR 33 per ton, though the amount of aid was dependent on the quantity of aid in respect of which aid was sought not exceeding a maximum guaranteed quantity. Article 89 laid down a maximum guaranteed quantity per marketing year which was to be apportioned among the Member States concerned as national guaranteed quantities in accordance with point B of Annex XI of the Single CMO Regulation. If more aid was sought than that laid down in the national guaranteed quantity, under Article 88(2) the aid was reduced in the Member State where production exceeded the national guaranteed quantity. Aid could first be paid when the dried fodder left the processing plant. Under Article 87, processing undertakings could, under certain conditions, be entitled to an advance payment of the aid.

As stated above, the first of the current aid schemes for the processing of an agricultural product was for the processing of flax and hemp. Article 91 provides that aid is paid for the processing of long and short straw flax and hemp grown for fibre production. The aid is granted only to authorized primary processors on the basis of the quantity of fibre actually obtained from straw for which a contract of sale has been concluded with a farmer.[90] Where a farmer retains ownership of the straw which is processed under contract by an authorized primary processor and the farmer can prove that they have placed the fibres obtained on the market, the aid is granted to the farmer. In situations where the authorized primary processor and the farmer are one and the

---

87. These rules were repealed by Art. 4(17) and Art. 8(e) of Council Regulation (EC) No 72/2009.
88. See recital 14 and Art. 8 of Council Regulation (EC) No 72/2009.
89. Under Art. 90 the Commission had authority to determine implementing provisions in the form of requirements for documentation from undertakings.
90. On the basis of Art. 95, the Commission has adopted implementing provisions in Commission Regulation (EC) No 507/2008. The term 'authorized primary processor' is defined in Art. 91(2).

same person, the contract of sale is replaced by a commitment by the party concerned to carry out the processing itself. As with the aid scheme for dried fodder, there are maximum guaranteed quantities for each marketing year, and these are allocated between the Member States pursuant to point A, II, of Annex XI of the Single CMO Regulation. According to Article 94(3), third paragraph, aid cannot be paid in excess of the national guaranteed quantity of the Member State.

The aid scheme for the processing of flax and hemp only applied for marketing years 2009/2010 to 2011/2012. Originally, there was no time limit, but the Union legislator inserted this limit in Council Regulation (EC) No 72/2009. As with the termination of grants of processing aid for drying fodder, the reason for this was partly the continued shift from production support to producer support by abolishing the existing aids and integrating support for these products into the system of decoupled income support for each farm. The Union legislator recognized that, even if the decoupling of aid to farmers resulted in the payment of the same amounts, it would considerably increase the effectiveness of the income support. In order to give the flax and hemp sector the possibility of adjusting to the removal of processing support, and to take account of the legitimate expectations of producers, the single payment scheme was introduced in a transitional period which was in force until 1 July 2012.[91]

For the second of the current organizations for processing aid, Article 95a(1) provides that potato starch producers are paid a premium of EUR 22.25 per tonne of starch produced. The premium is paid for the quantity of potato starch up to the quota limit referred to in Article 84a(2) of the Single CMO Regulation.[92] It is a condition for the payment of the premium that the starch producer has paid the potato producers a minimum price for all the potatoes necessary to produce starch up to that quota limit. Under Article 95a(2), the minimum price of potatoes intended for the manufacture of potato starch is EUR 178.31 per tonne. This price applies to the quantity of potatoes delivered to the factory which is needed to make one tonne of starch. Article 95a(2) gives authority to adjust the minimum price according to the starch content of the potatoes.[93]

[2] Production Refunds

The second form of aid scheme for certain agricultural products involves production refunds. The purpose of this form of aid is to compensate for the lack of or the limited production of certain products within the Union. According to Article 97 of the Single CMO Regulation, production refunds may be granted on certain products of the sugar

---

91. Recitals 12 and 13 of Council Regulation (EC) No 72/2009. See Ch. 3, §3.02[D], on the fundamental principle of the protection of legitimate expectations.
92. On market intervention in the form of production regulation and quota schemes for potato starch, see §4.04[C][3] above.
93. Under the authority of Art. 95a(3), Commission Regulation (EC) No 571/2009 lays down detailed provisions for the adjustment of minimum prices.

sector.[94] Under Article 98, the Commission can adopt the conditions for the granting of production refunds, the amount of such refunds and the eligible quantities.

[3]   Aid for Milk and Milk Products

Aid for skimmed milk and skimmed milk powder falls under the headings of three different schemes. Under the first scheme, aid is granted for skimmed milk and skimmed milk powder and buttermilk and buttermilk powder intended for use as feeding stuffs. It is a condition for such aid that there should be a surplus of milk products, or a risk of a surplus, which creates or could create a serious imbalance in the market. The Commission can decide to grant aid in these situations, in which case Article 99(1) authorizes the Commission to determine the conditions and product standards to be applied. The amount of aid can be set by the Commission in advance within certain limits or by tendering.[95]

The second scheme and group of products for which aid is granted concerns skimmed milk processed into casein and caseinates (milk protein). Under Article 100, it is the Commission that lays down the conditions and product standards for the aid, and within certain limits the Commission also determines the amount of the aid.[96] Under this scheme, it is a fundamental condition for the payment of aid that there should be a surplus of milk products, or a risk of such surplus, which creates or could create a serious imbalance in the market.

The third of the existing aid schemes is provided for in Article 102(1), whereby aid is granted for supplying certain processed milk products to pupils in educational establishments. The group of such products is determined by the Commission within certain guidelines.[97] In addition to Union aid, Member States may grant national aid for supplying the products referred to in paragraph 1 to pupils in educational establishments. Member States may finance their national aid by means of a levy on the dairy sector or by any other contribution from the dairy sector.[98] The provision should presumably be understood as excluding the use of other forms of financing by individual Member States.

Until 1 July 2009, a rule in Article 101 applied under which aid could be given for the purchase of cream, butter and concentrated butter at reduced prices. The Commission could give such aid with a view to mitigating the effects of over-production of milk

---

94. Up until 30 June 2009, there were also production refunds for maize, wheat and potato starch, as well as certain derivative products; see Art. 4(23) and Art. 8(b) of Council Regulation (EC) No 72/2009.
95. See Commission Regulation (EC) No 2799/1999 laying down detailed rules for applying Regulation (EC) No 1255/1999 as regards the grant of aid for skimmed milk and skimmed milk powder intended for animal feed, and the sale of such skimmed milk powder.
96. See Commission Regulation (EC) No 1487/2006 amending Regulation (EEC) No 2921/90 as regards the amount of the aid for the production of casein and caseinates from skimmed milk.
97. See Commission Regulation (EC) No 657/2008 of 10 Jul. 2008 laying down detailed rules for applying Council Regulation (EC) No 1234/2007 as regards Union aid for supplying milk and certain milk products to pupils in educational establishments.
98. Article 183 of the Single CMO Regulation gives Member States authority to impose a promotional levy on the milk and milk products sector.

products. The provision listed those groups that could be offered the products at reduced prices. These were: non-profit making institutions and organizations; military forces and units of comparable status in the Member States; manufacturers of pastry products and ice-cream; manufacturers of other foodstuffs to be determined by the Commission; and for the direct consumption of concentrated butter.[99] Article 101 was repealed by Council Regulation (EC) No 72/2009. The reason for this was that, in line with the reduction of the intervention price for butter, ultimately to zero, aid for the sale of the products covered by the aid scheme was also reduced to zero. The Union legislator therefore found that, in the light of the favourable market situation, the aid for sales was no longer necessary and should therefore be abolished.[100]

[4] Aids in the Hops Sector

With effect from 1 January 2011, Council Regulation (EC) No 72/2009 introduced a new Article 102a into the Single CMO Regulation, with special provisions on aid to the hops sector.[101] The background to this was that Council Regulation (EC) No 1782/2003 establishing common rules for direct support schemes under the common agricultural policy contained a rule whereby Member States could retain part of the component of national ceilings corresponding to the hops area payments and use them, in particular, to finance certain activities of recognized producer organizations.[102] Since Regulation (EC) No 1782/2003 was repealed by Regulation (EC) No 72/2009, and area payments for hops were decoupled, this meant that at the same time payments to producer organizations were terminated. The aim of Article 102a is for equivalent amounts to be used in the Member State concerned for the same activities in order to allow the hop producer organizations to continue their activities as before.[103]

[5] Aids in the Olive Oil and Table Olives Sector

Under the provision of Article 103 of the Single CMO Regulation, in the olive oil and table olives sector aid is paid for three-year work programmes to be drawn up by operator organizations. These work programmes can include, for example: market follow-up and administrative management in the olive oil and table olives sector; improvement of the production quality of olive oil and table olives; or dissemination of information on the activities carried out by operator organizations with the aim of improving the quality of olive oil. The term 'operator organizations' is used in the sense set out in Article 125, i.e., recognized producer organizations, recognized inter-branch organizations or recognized organizations of other operators in the olive oil and table

---

99. 'Manufacturers of other foodstuffs' were defined in Commission Regulation (EC) No 1898/2005 of 9 Nov. 2005 laying down detailed rules for implementing Council Regulation (EC) No 1255/1999 as regards measures for the disposal of cream, butter and concentrated butter on the Community market.
100. See recital 11, Art. 4(25) and Art. 8(b) of Council Regulation (EC) No 72/2009.
101. See Art. 4(27) and Art. 8(d) of Council Regulation (EC) No 72/2009.
102. See §4.05[B] below on private organizations and associations as part of a market organization.
103. See recital 18 of Council Regulation (EC) No 72/2009.

olives sector or their associations.[104] Article 103(1a) provides that the annual Union financing of the work programmes shall be: EUR 11,098,000 for Greece, EUR 576,000 for France, and EUR 35,991,000 for Italy. There are also a number of detailed provisions on the amount of aid and its financing.

[6]   *Aid for the Fruit and Vegetables Sector*[105]

Aid for the fruit and vegetables sector is divided into three schemes: aid to producer organizations, aid for setting up operational funds and operational programmes, and aid for schemes for the free distribution of fruit to schools etc.

Producers of fresh fruit and vegetables may join together in producer organizations with the aim of supporting environmentally sound production, administration and growing methods. Such organizations can also contribute to the rationalization of production and the marketing of products. Among other things, membership of a producer organization means that producers bind themselves to comply with the organization's rules for production, sale and protection of the environment, and that they contribute to the organization's operational fund.[106]

According to Article 103a, together with Article 125e, aid for producer organizations concerns organizations in the Member States which joined the European Union on 1 May 2004 or subsequently, in the outermost regions of the Union, or in the smaller Aegean Islands. Article 103a(1) states generally that the Member States can provide aid, but at the same time that this can only be during 'the transitional period allowed pursuant to Article 125e'; Article 125e relates only to the Member States and territories referred to here. The geographical restriction of the aid to producer organizations is also clear from Article 103a(3), which lays down the limits for the amount of the aid, specifying the same areas. The aid can be given in general to promote the setting up of such organizations.

According to Article 103b, producer organizations in the fruit and vegetables sector may set up an operational fund, financed by the financial contributions of members or of the producer organization itself and by Union financial assistance. It is a condition for the grant of such aid that the operational funds are only used to finance operational programmes that have been approved by Member States. Article 103g lays

---

104. Prior to 2009, Art. 103(1) provided that the aid should be financed by amounts withheld by Member States in accordance with special rules for the olives sector. Article 103(1) was amended by Art. 4(28) of Council Regulation (EC) No 72/2009. On producer and inter-branch organizations in the olive oil and table olives sector, see §4.05[B] below.

105. Alongside the negotiations for the adoption of the Single CMO Regulation, the Council also carried on negotiations on and adopted a reform of the fruit and vegetables sector and the processed fruit and vegetables sector. Council Regulation (EC) No 1182/2007 of 26 Sep. 2007 laying down specific rules as regards the fruit and vegetables sector was adopted in this connection. Recital 8 to the Single CMO Regulation explained that only provisions on fruit and vegetables and processed fruit and vegetables should not be included from the start, and that substantive provisions should first be included when the respective reforms had been made. The fruit and vegetables and processed fruit and vegetables sectors were first fully incorporated in the Single CMO Regulation by Council Regulation (EC) No 361/2008.

106. See §4.05[B] below for further on producer organizations.

down a number of formal requirements that must be fulfilled in connection with the approval of operational programmes by national authorities, and Article 103c lays down substantive requirements for operational programmes. Such programmes must either pursue two or more of the aims listed in Article 122(c) or two or more of the aims listed in Article 103c(1).

Article 122(c) lists three aims for producer organizations in the fruit and vegetables sector: ensuring that production is planned and adjusted to demand, particularly in terms of quality and quantity; concentrating supply and the placing on the market of the products produced by its members; optimizing production costs and stabilizing producer prices.

The other list of possible aims for producer organizations in Article 103c(1) lists six possible aims. These are: planning of production; improvement of product quality; boosting products' commercial value; promoting the products, whether in a fresh or processed form; environmental measures and methods of production respecting the environment, including organic farming; crisis prevention and management. What is meant by crisis prevention and management is set out in Article 103c(2).

According to the wording of Article 103c, the fulfilment of the requirement for a minimum number of aims cannot be achieved by combining one of the aims listed in Article 103c with one of the aims listed in Article 122(c). However, this is unlikely to have any practical effect, since there is some overlap between the aims listed in both these provisions.

As a further requirement for operational programmes, Article 103c(3) provides that the Member States must ensure that operational programmes include two or more environmental actions, or that at least 10% of the expenditure of operational pro-grammes covers environmental actions.[107] Article 103c(5) adds that investments which increase environmental pressure are only permitted if there are safeguards in place to protect the environment from these pressures.

Article 103d provides that Union financial assistance must be limited to 50% of the actual expenditure incurred, and must be capped at 4.1% of the value of the marketed production of each producer organization. Under certain conditions, these limits can be raised to 60% (and in some situations even 100%) and 4.6% respectively. Under Article 103e, where the degree of organization of producers in the fruit and vegetables sector is particularly low, Member States may be authorized by the Commission to pay national financial assistance to producer organizations. Article 103e previously had a paragraph (2), according to which, pursuant to Article 42 TFEU, the Treaty rules on State aid did not apply to State aid that was approved pursuant to paragraph (1) of the Article. This provision was moved to Article 180 of the Single CMO Regulation by Council Regulation (EC) No 72/2009.[108]

---

107. Under Art. 103f, the Member States must establish a national strategy for sustainable operational programmes in the fruit and vegetables market.
108. See Art. 3(34) of Council Regulation (EC) No 72/2009.

Article 103h authorizes the Commission to establish detailed rules for the application of aid arrangements for the fruit and vegetables sector.[109]

[7]   Aid for School Fruit Schemes

The rules on aid for school fruit schemes, in the form of aid for the supply of fruit and vegetables, processed fruit and vegetables and banana products to children, are to be found in Article 103ga. This provision was inserted in the Single CMO Regulation by Council Regulation (EC) No 13/2009, in which the Union legislator stated that it was 'desirable to address the low consumption of fruit and vegetables amongst children'. The measure involves ensuring a durable increase of the share of fruit and vegetables in the diets of children at the stage when their eating habits are being formed. It is also stated in the preamble to Regulation (EC) No 13/2009 that the provision of Union aid under a school fruit scheme to supply fruit, vegetable and banana products to children in educational establishments should bring young consumers to appreciate fruit and vegetables and thus enhance future consumption.[110]

According to Article 103ga(1)(a), the target group of the school fruit scheme is children in educational establishments, including nurseries, other pre-school establishments, primary and secondary schools.[111] Article 103ga(1)(a) and (b) states that aid can be given for the supply of fruit and vegetable products to children in educational establishments, as well as certain related costs of logistics and distribution, equipment, communication, monitoring and evaluation. Article 103ga(4)(c) expressly states that aid cannot cover other costs.

It is up to each Member State to decide whether it wants to participate in the school fruit scheme. If it decides to participate, Article 103ga(2) requires the Member State to draw up a national or regional plan for implementing the scheme. The provision specifies that the strategy must include the budget of the scheme, its duration, the target group, the eligible products and the involvement of relevant stakeholders. According to Article 103ga(3), Member States must draw up a list of products of the fruit and vegetables, processed fruit and vegetables, and bananas sectors that will be eligible under their schemes. This list of products must be selected on the basis of objective criteria which may include seasonality, availability of produce or environmental concerns. On the basis of this wording, it must be assumed that Member States may take account of other objective criteria. Article 18 TFEU expressly

---

109. See Commission Implementing Regulation (EU) No 543/2011 which replaced Commission Regulation (EC) No 1580/2007.
110. See Council Regulation (EC) No 13/2009, recital 2. In the same place it is also stated that the school fruit scheme will meet the objectives of the common agricultural policy, including the promotion of earnings in agriculture, the stabilization of markets and the availability of both current and future supplies. On the statement of the objectives of the common agricultural policy in Art. 39(1)(b), (c) and (d), see Ch. 2, §2.05[B][2]–§2.05[B][4].
111. According to Art. 103ga(5), the aid is allocated to each Member State on the basis of their proportion of six to ten year old children.

prohibits discrimination on the ground of nationality.[112] However, Union law does not contain any general prohibition of discrimination between agricultural products from the Union and agricultural products from third countries. The Union may have bound itself not to discriminate in agreements with third countries or other international organizations, but where this is not the case, Article 103ga(3) expressly states that, in choosing which products may be included in a school fruit scheme, Member States may give preference to products of Union origin.[113]

The Union legislator has laid down a limit to the total amount that may be spent in support of school fruit schemes in the Member States. Under Article 103ga(4), the total Union aid may not exceed EUR 90 million per school year, or 50% of the costs of supply and related costs.[114]

It is the Commission that allocates the aid for school fruit schemes between the Member States. Each year, participating Member States apply for Union aid on the basis of their strategies. The Commission then decides on definitive allocations, within the appropriations available in the budget. Allocation is made on the basis of objective criteria based on the Member State's proportion of 6 to 10 year old children. However, the provision states that participating Member States each receive aid of at least EUR 175,000. This fixed amount clearly favours smaller Member States.

It is possible for individual Member States to have national school fruit schemes alongside or in combination with a Union aid scheme. The rules on this are in Article 103ga(6) to (8).

Article 103ga(1) and 103h(f) authorize the Commission to lay down more detailed implementing provisions, and it has exercised this power in Commission Implementing Regulation (EU) No 1208/2011.

[8] Support for the Wine Sector

The wine sector was fully included under the Single CMO Regulation by Council Regulation (EC) No 491/2009. This also applies to the rules on support programmes for the sector.[115] These rules are now in Article 103i–103za. The first subsection (Article 103i and 103j) contains introductory provisions. Thereafter, there is a subsection (Article 103k–103n) which lays down general conditions for support programmes, and a third subsection (Article 103o–103z) which contains provisions for specific support measures. In Article 103za, there is a provision whereby the Commission is authorized to adopt measures necessary for the implementation of support programmes for the wine sector.[116]

The basic system for the support programmes in the wine sector is set out in Article 103i. The regulatory system covers both the grant of Union funds to the Member

---

112. See Ch. 3, §3.02[B], on the prohibition of discrimination on the ground of nationality as an example of the fundamental principle of equal treatment in Union law.
113. On the EU's relations with other international organizations, including the WTO, see Ch. 9.
114. The ceiling is 75% in a special group of regions in the Union; see Art. 103ga(4)(b).
115. See recital 3 of Council Regulation (EC) No 491/2009.
116. The Commission has exercised this power in Commission Regulation (EU) No 772/2010.

States and the Member States' use of these funds in connection with national support programmes for the financing of specific support measures in the wine sector. In other words, it is the individual Member State that develops the individual support programme, and it is the Union that finances the programme. However, the Union's expenditure on the support programmes cannot be allowed to run out of control. Article 103n(1) and Annex Xb to the Regulation set out the total Union funds that can be used on the support programmes and the allocation of these between the Member States. According to Article 103n(3), in principle the Member States may not contribute to the costs of measures financed by the Union under the support programmes. However, Article 103n(4) modifies this by allowing Member States to derogate from paragraph 3 by granting national aid to Union-financed support measures for promotion on third-country markets, harvest insurance and investments.[117] Article 103n(4) also provides that the 'maximum aid rate as laid down in the relevant [Union] rules on State aids' applies to the global public financing, including both Union and national funds. The rule was included in the Single CMO Regulation by Council Regulation (EC) No 491/2009, where it is emphasized that any element of State aid included in the national support programmes must be assessed in the light of the Union's substantive rules on State aid.[118]

Article 103j(1) states that support programmes must be compatible with Union law and consistent with the activities, policies and priorities of the Union. Article 103j(2) makes clear the responsibilities of Member States. Individual Member States are responsible for their support programmes and for ensuring that they are internally consistent and drawn up and implemented objectively. In developing a support programme, the Member State must take into account the economic situation of the producers concerned and the need to avoid unjustified unequal treatment between producers. This provision on the equal treatment of wine producers is an expression of the general Union principle of equal treatment.[119] Each Member State is not only responsible for the content and implementation of individual support programmes, Article 103j(2) stipulates that they are also responsible for providing for and carrying out the necessary controls and penalties in case of non-compliance with the support programmes. Under Article 103j(3), no support may be granted for research projects or for measures contained in rural development programmes.

The fact that the Single CMO Regulation leaves it to individual Member States to draw up their own support programmes in the wine sector allows for greater flexibility and thus makes it easier to adjust specific support to national, regional and local circumstances. However, this method for regulating support programmes also involves a risk of there being inappropriate developments and even abuse. In addition to the rules in Article 103j(2) and (3) referred to, which emphasize the Member States' responsibilities for the support programmes, Article 103k–103n therefore establish

---

117. Article 103n(4) refers to the rules on State aid for the measures referred to in Arts. 103p, 103t and 103u.
118. Recital 7 of Council Regulation (EC) No 491/2009. See Ch. 8, §8.04, on State aid for undertakings in the agricultural sector.
119. On the principle of equal treatment, see Ch. 3, §3.02[B].

more formal rules for the submission of support programmes to the Commission and for the content of programmes. Only the 18 producer Member States listed in Annex Xb of the Single CMO Regulation are subject to the rules on the submission of support programmes to the Commission. Under Article 103k(1), these producer Member States must submit to the Commission a draft five-year support programme. Each Member State must submit a single draft support programme which may accommodate regional particularities.

There is no provision which expressly states that the Commission must approve the support programmes of the Member States, but this is the reality of Article 103k. As stated, Article 103k(1) requires the Member States to submit their programmes to the Commission. Under Article 103k(2), support programmes become applicable three months after their submission to the Commission. The Commission uses those three months to assess whether the programme fulfils the conditions of the Single CMO Regulation. If the Commission finds that a submitted support programme does not comply with the conditions, it must inform the Member State thereof. In such a case, Article 103k(2), second paragraph, requires the Member State to submit a revised support programme to the Commission. Article 103k(2) must thus be interpreted as meaning that Member States may not implement support measures contained in their support programmes before the expiry of the three months deadline, and that if a Member State has not heard from the Commission before the expiry of the deadline, it must be entitled to proceed without receiving the opinion of the Commission. If a Member State is required to submit a revised support programme to the Commission pursuant to Article 103k(2), this can be initiated by the State two months after submission. If, within this two-month period, the Commission nevertheless finds that the support programme does not fulfil the conditions laid down, the Member State must refrain from initiating it and once again submit an amended programme to the Commission.

If a Member State wishes to amend a support programme which the Commission has already approved, then according to Article 103k(3), it must follow the same procedure as for submitting a new support programme to the Commission.

Prior to submitting a support programme to the Commission, the Member State concerned must have consulted the competent authorities and organizations at the appropriate territorial level. Article 103k(1), third paragraph, leaves it to the Member State to draw up support measures in their support programmes which it deems most appropriate at the geographical level, but the consultations must take place at the relevant geographic level.

Support programmes submitted to the Commission by Member States must satisfy the seven requirements as to their content, as set out in Article 103l. Support programmes must contain a detailed description of the measures proposed as well as their quantified objectives. Each Member State has only one and may only submit one support programme, which must contain one or more of the measures eligible for support listed in Article 103m.[120] Next, according to Article 103l, a support programme

---

120. See immediately following for the 11 measures eligible for support.

must contain the results of consultations held by the Member State. It must include an appraisal of the expected technical, economic, environmental and social impact of the support programme. As stated in Article 103k(1), the support programme must be for a five-year period, but the support measures which it contains need not be. Thus, as a fourth requirement under Article 103l, the programme must contain a schedule for implementing the measures. As the fifth requirement, the support programme must contain a general financing overview showing the resources to be deployed. This overview must also show the envisaged indicative allocation of the resources between the measures. The financial overview must show that the financing of the support programme complies with the budgetary ceilings laid down in Annex Xb to the Regulation, pursuant to Article 104n(1). As stated above, the rules on support programmes in the wine sector have special requirements for following up their implementation. Thus, the sixth requirement is that the support programme must contain the criteria and quantitative indicators to be used for monitoring and evaluation, as well as the steps taken to ensure that the support programmes are implemented appropriately and effectively. Finally, the support programme must designate the competent authorities and bodies of the Member State responsible for implementing the support programme.

The measures eligible for support that can be included in the support programme of an individual Member State are exhaustively listed in Article 103m(1). The provision lists 11 specific support measures which are described in more detail in Article 103o–103y.[121] It is expressly stated in Article 103m(2) that support programmes may not contain measures other than those listed.

The first kind of measure which may be included in a support programme, in Article 103m(1)(a), is support for a single payment scheme in accordance with Article 103o. Article 103o(1) gives Member States the right to grant support to vine growers by allocating to them payment entitlements pursuant to Regulation (EC) No 1782/2003. If a Member State intends to grant such support, then according to Article 103o(3)(a), it must be part of a single payment scheme. A Member State may not abandon a decision, once taken, to use a single payment scheme as a support measure, nor can this be achieved by way of changes to support programmes in accordance with Article 103k(3). Under Article 103o(3)(b), the use of a single payment scheme in the wine sector will commensurately reduce the amount of funds available for support measures listed in Article 103p–103y.

The second kind of measure which a Member State can include in its support programme is support for sales promotion to third-country markets. Under Article 103p(1), such support must cover information or promotion measures concerning Union wines in third countries. The measure must aim to improve the competitiveness of wines in the countries concerned. Article 103p(3) specifies which kinds of measures these can be. The list is exhaustive but broad, as it includes studies of new markets,

---

121. Support measures referred to in Art. 103w (potable alcohol distillation), Art. 103x (crisis distillation), and Art. 103y (use of concentrated grape must), could be included in support programmes up to 31 Jul. 2012. See immediately below.

sales promotion and advertising, participation at fairs or exhibitions, and information campaigns. According to Article 103p(4), the Union's contribution to promotion activities may not exceed 50% of the eligible expenditure. Under Article 103p(2), only wines with a protected designation of origin or a protected geographical indication or wines with an indication of the wine grape variety can benefit from support measures.[122]

The third kind of measure that is eligible for support is measures relating to the restructuring and conversion of vineyards. According to Article 103q(1), the overall aim is to increase the competitiveness of wine producers, and such measures can only cover three specific activities. The normal renewal of vineyards which have come to the end of their natural life may not be supported. Support can only be given for varietal conversion, relocation of vineyards, and improvements to vineyard management techniques. Article 103q sets an upper limit to the Union's contribution, which may not exceed 50% of the actual costs of the restructuring and conversion of vineyards.[123]

'Green harvesting' is the fourth kind of support measure which may be included in a Member State's support programme. Article 103r(1) defines 'green harvesting' as the total destruction or removal of grape bunches while still in their immature stage, thereby reducing the yield of the relevant area to zero. This support must have the aim of contributing to restoring the balance between supply and demand on the Union wine market and thus to preventing a collapse of the market. According to Article 103r(3), support for green harvesting may be granted in the form of a flat rate payment per-hectare to be determined by the Member State concerned. The payment may not exceed 50% of the direct costs of the destruction or removal of grape bunches and the loss of revenue related to such destruction or removal. The importance of this limit being complied with by the Member States is emphasized in Article 103r(4). This rule binds the Member States concerned to establish a system based on objective criteria to ensure that green harvesting measures do not lead to compensation of individual wine producers in excess of the ceiling referred to.

Article 103s lays down the provisions for the fifth support measure in the form of the establishment of mutual funds. These mutual funds must be for the purpose of assisting producers who seek to insure themselves against market fluctuations. The support may take the form of temporary and degressive aid to cover the administrative costs of the funds.

As the sixth form of support, under Article 103t Member States can provide support for harvest insurance. Support for harvest insurance may be granted in the form of a financial contribution from the Union. Harvest insurance must fulfil several conditions in order to be considered a measure entitled to support. First, the harvest insurance must be intended to contribute to safeguarding producers' incomes where these are affected by natural disasters, adverse climatic events, diseases or pest infestations. Second, support for harvest insurance may only be granted if the insurance payments do not cover more than 100% of the income loss suffered. In

---

122. See §4.05[A][10] on designations of origin and protected geographical indications in the viticulture sector.
123. There is a higher ceiling of 75% for regions classified as convergence regions.

calculating the level of cover, account must be taken of any compensation the producers may obtain from other support schemes related to the insured risk. Finally, support for harvest insurance may not distort competition in the insurance market.[124] There are limits to how large a proportion of the producer's costs the support may cover. According to Article 103t(2), the support may cover 80% of the cost of the insurance premiums for insurance against losses resulting from adverse climatic events which can be categorized as natural disasters. Support may only cover 50% of the cost of the insurance premiums for insurance against losses caused by adverse climatic events other than natural disasters and losses caused by animals, plant diseases or pest infestations.

The seventh kind of measures in respect of which Member States are entitled to provide support are tangible or intangible investments in processing facilities, winery infrastructure and marketing of wine which improve the overall performance of the enterprise.[125] It is a condition that the investment must relate to one or more expressly named items. Article 103u(1)(a) refers to the production or marketing of products referred to in Annex XIb of the Single CMO Regulation. The Annex lists and defines different categories of grapevine products, including: wine, new wine still in fermentation, liqueur wine, various kinds of sparkling wine, semi-sparkling and aerated wine, various kinds of grape must, wine from raisined grapes, wine from overripe grapes and wine vinegar. According to Article 103u(1)(b), the support can also be for the development of new products, processes and technologies related to the products referred to in Annex XIb.

Support for investments may not cover the total costs. Article 103u(4) lays down some maximum rates for how large a share of an individual investment may be covered by support. The provision has various maximum rates based on geographical criteria. For example, there is a maximum contribution of 50% in regions classified as convergence regions in accordance with Regulation (EC) No 1083/2006. Based on its wording, it must be assumed that undertakings that are not covered by the provision may not be granted investment support.

Article 103u(2) contains special provisions for certain categories of undertakings that are covered by paragraph (1). The provision refers to three categories of undertakings, and the wording must be interpreted as meaning that undertakings that fall within these categories are automatically covered by the provision in paragraph (4). The first of these categories is micro, small and medium-sized enterprises.[126] This group can receive support up to the maximum rate given in Article 103u(4). The second

---

124. On the rules on the distortion of competition in the agricultural sector, see §4.07 below, and Ch. 8, §8.03.
125. Article 103u on support for investment was not included in the Commission proposal (COM(2007) 372 final) which led to the adoption of the common market organization for wine in Council Regulation (EC) No 479/2008, the provisions of which, particularly Art. 15, correspond with few alterations to Art. 103u, as subsequently incorporated in the Single CMO Regulation. The provision on support for investments was presumably included in connection with the Council procedure, and there are no published *travaux préparatoires* for the provision.
126. For the definition of these kinds of undertaking, see Art. 103u(2), which refers to Commission Recommendation 2003/361/EC concerning the definition of micro, small and medium-sized enterprises.

group consists of undertakings located in the Azores, Madeira, the Canary Islands, the smaller Aegean Islands and the French overseas departments. For this group there are no size limits for applying the maximum rate. The third group of undertakings for which a special maximum rate applies, is undertakings which employ 250 or more employees and have an annual turnover of over EUR 50 million, or an annual balance of over EUR 43 million,[127] and have fewer than 750 employees or a turnover of less than EUR 200 million. For this group of undertakings, investment support can be given for up to half the maximum rate. Presumably, Article 103u(2) must be interpreted so that if an undertaking is not covered by a provision it cannot be covered by an investment programme even if the undertaking is covered by Article 103u(4).

According to Article 103u(2), support for investments may not be granted to enterprises in difficulty within the meaning of the Union guidelines on State aid for rescuing and restructuring firms in difficulty.[128] Moreover, under Article 103u(3), costs eligible for support do not include VAT, except non-recoverable VAT when it is genuinely and definitively borne by beneficiaries other than non taxable persons, interest on debt, or the purchase of land costing more than 10% of the support.[129]

Article 103v, 103w, and 103x contain rules for the eighth, ninth and tenth kinds of measure in respect of which Member States are entitled to provide support, in connection with various forms of distillation of wine. Article 103v concerns by-product distillation. Individual Member States may include in their support programmes support measures for the voluntary or obligatory distillation of by-products of wine-making. The by-products must meet the conditions laid down in point D of Annex XVb of the Single CMO Regulation. The maximum aid measures are laid down by the Commission and must be based on collection and processing costs.[130] Article 103v(3) provides that the alcohol resulting from the supported distillation must be used exclusively for industrial or energy purposes so as to avoid distortion of competition.[131]

Article 103w and 103x are provisions which allowed support measures until 31 July 2012. Under Article 103w, up to that date support could be paid to producers for the distillation of wine into potable alcohol. In this case the support was to be paid in the form of per-hectare aid. Under Article 103x, on crisis distillation, support could be granted for voluntary or obligatory distillation of surplus wine where the individual Member State wanted to reduce or eliminate a surplus and at the same time ensure supply continuity from one harvest to the next.

---

127. In connection with this special group of undertakings, Art. 103u(2) refers to Commission Recommendation 2003/361/EC concerning the definition of micro, small and medium-sized enterprises, Annex I, section I, point 2(1).
128. See Ch. 8, §8.04, on State aid in the agricultural sector.
129. Article 103u(3) refers to Art. 71(3)(a)-(c) of Council Regulation (EC) No 1698/2005 on support for rural development by the European Agricultural Fund for Rural Development. On the Rural Development Regulation, see Ch. 5.
130. For an older example of the Commission's approval of a Member State's support plan for distillation of wine, see Commission Decision 93/155/EEC concerning an aid measure proposed by the German authorities (Rhineland-Palatinate) for the distillation of wine (OJ L 61, 13 Mar. 1993, 55).
131. See Ch. 8, §8.04, on State aid to undertakings in the agricultural sector.

As the final support measure that can be included in Member States' support programmes in the wine sector, Article 103y provided that support could be granted to wine producers who used concentrated grape must to increase the natural alcoholic strength of products. The scope for using this measure also came to an end on 31 July 2012.

At the same time as the wine sector was fully incorporated under the Single CMO Regulation by Council Regulation (EC) No 491/2009, Article 103z on cross compliance was inserted in the Regulation. The requirement for cross compliance applies to support measures for restructuring and conversion pursuant to Article 103q, and to green harvesting pursuant to Article 103r. The legislative and administrative requirements and conditions for good agricultural and environmental practice must be complied with within three years in respect of restructuring and conversion, and one year in respect of green harvesting.[132] Article 103z refers to Articles 3–7 of Regulation (EC) No 1782/2003; the relevant provisions are now in Articles 4–6 and 23 and 24 of Council Regulation (EC) No 73/2009.[133]

Article 103za authorizes the Commission to adopt more detailed implementing measures in respect of the Single CMO Regulation's provisions on support programmes for the wine sector, and it has exercised these powers in Commission Regulation (EC) No 555/2008.

[9]   Aid for Raw Tobacco

Council Regulation (EEC) No 2075/92 on the common organization of the market in raw tobacco originally established a premium system in order to help secure the incomes of producers of raw tobacco.[134] The premium system was in force up to and including the harvest of 2005, after which up to and including 2009 support was paid pursuant to the rules in Council Regulation (EC) No 864/2004. The most important part of the original market organization for raw tobacco, which is carried forward in the Single CMO Regulation, is Article 104 on the Union Tobacco Fund. The Fund finances two kinds of measures. On the one hand, it finances measures to promote the awareness of citizens of the harmful effects of the use of tobacco in any form. This is done through the dissemination of information and through education, as well as through scientific initiatives such as the collection of information about trends in tobacco use and carrying out epidemiological studies into the use of tobacco. The

---

132. The Commission proposed a deadline not of three years but of five years; see COM(2007) 372 final, 29. In its response to consultation pursuant to Art. 37 of the EU Treaty, the European Parliament proposed that the provision on cross compliance should be dropped entirely from the Regulation. The response to consultation did not give any reason, but the proposal may be due to the fact that elsewhere the European Parliament has suggested that the concept of cross compliance is associated with direct payment, so that in the view of the Parliament, a regulation on a market organization ought only to refer to the principles and norms associated with cross compliance; see the European Parliament's Report of 28 Nov. 2007, A6-0477/2007, pr. 239 and 87, on the proposal for a Council regulation on the common organization of the market in wine and amending certain Regulations.
133. See Council Regulation (EC) No 73/2009, Annex XVIII.
134. Greece and Italy are by far the largest producers of raw tobacco in the EU.

second group of measures is used to help tobacco growers switch to other crops or other economic activities.

The Fund itself was originally financed by retaining a certain percentage of the premium that was paid to tobacco producers, while pursuant to Article 104(2)(b) of the Single CMO Regulation, up to the end of calendar year 2009 financing was in accordance with Article 110m of Council Regulation (EC) No 1782/2003.[135] For calendar years 2008 and 2009, the financing consisted of 5% of the support granted to the producers of raw tobacco.[136]

[10] Aid for the Apiculture Sector

Article 105 of the Single CMO Regulation gives individual Member States the possibility of drawing-up a national apiculture programme for a period of three years.[137] The main purpose of an apiculture programme must be the improvement of the general conditions for the production and marketing of apiculture products and it can include a number of the measures listed in Article 106. Article 106 contains an exhaustive list of the measures which may be included in an apiculture programme, as follows: technical assistance to beekeepers and groupings of beekeepers; control of varroasis; rationalization of transhumance; measures to support laboratories carrying out analyses of the physico-chemical properties of honey; measures to support the restocking of hives in the Union; and cooperation with specialized bodies for the implementation of applied research programmes in the field of beekeeping and apiculture products.

If a measure is financed by the EAFRD in accordance with Council Regulation (EC) No 1698/2005, it is not a measure entitled to aid in apiculture programme under Article 106.[138]

According to Article 109, an apiculture programme must be drawn up in close collaboration with the representative organizations and beekeeping cooperatives and approved by the Commission. Under Article 108, the Union then provides part-financing for apiculture programmes equivalent to 50% of the expenditure borne by the Member States.[139] Previously, Article 105(2) of the Single CMO Regulation contained a rule that both the financial contribution provided by Member States for measures subject to Union support and specific national aids were exempt from the

---

135. Article 104(2)(b) was inserted in the Single CMO Regulation by Council Regulation (EC) No 470/2008. It is not clear from the provision that Art. 100m was first included in Council Regulation (EC) No 1782/2003 by Art. 1(20) of Council Regulation (EC) No 864/2004, and that Art. 100m was later amended by Art. 1 of Regulation (EC) No 470/2008. Regulation (EC) No 1782/2003, including Art. 100m, was repealed by Art. 146(1) of Council Regulation (EC) No 73/2009.
136. The Commission adopted implementing measures for the Tobacco Fund in Commission Regulation (EC) No 2182/2002.
137. The system of national aid programmes is also known in the olive oil and table olives sector and the wine sector; see §4.04[D][5] and §4.04[D][8] above.
138. EAFRD stands for the European Agricultural Fund for Rural Development. See Ch. 5, §5.02, on EAFRD.
139. The Commission has adopted implementing provisions for the apiculture sector in Commission Regulation (EC) No 917/2004.

State aid rules in Articles 107–109 TFEU. This was amended by Council Regulation (EC) No 72/2009, so that Article 105(2) now provides that Member States may pay specific national aids for the protection of apiaries disadvantaged by structural or natural conditions or under economic development programmes. However, this does not apply to aid for production or trade. The Member States must notify the Commission about such aid when submitting their apiculture programme to the Commission. Council Regulation (EC) No 491/2009 makes it clear that Articles 107–109 TFEU do not apply to payments pursuant to Article 105 of the Single CMO Regulation.[140]

[11]   Aids in the Silkworm Sector

Article 111(1) establishes that aid is to be granted to producers of silkworms and for silkworm eggs reared within the Union. The aid is to be granted to silkworm rearers in respect of each box of silkworm eggs used, provided the boxes contain a minimum quantity of eggs and that the worms have been successfully reared. Pursuant to Article 111(3), the aid per box of silkworm eggs used is EUR 133.26.

Under Article 112, the Commission is required to adopt detailed rules for the application of aids in the silkworm sector covering, in particular, the minimum quantity of eggs referred to in Article 111(2).[141]

§4.05   RULES CONCERNING MARKETING AND PRODUCTION IN THE INTERNAL MARKET

Title II of Part II of the Single CMO Regulation is headed 'Rules concerning marketing and production'. The Title is divided into two chapters, one containing Articles 113–121, laying down marketing standards and conditions for production, and the other containing Articles 122–127, containing rules on producer organizations, interbranch organizations, and operator organizations.

[A]   Rules Concerning Marketing and Production

[1]   Marketing Standards: The General Rules

The chapter containing the rules on marketing and production is divided into six sections, of which the first (Articles 113–118) is headed 'Marketing standards'. Article 113 gives the Commission authority to lay down marketing standards for one or more products in five different sectors. This concerns marketing standards for olive oil,[142] fruit and vegetables, processed fruit and vegetables, bananas and live plants. Article

---

140. See Art. 4(30) of Council Regulation (EC) No 72/2009, and Art. 1(26) of Council Regulation (EC) No 491/2009.
141. See Commission Regulation (EC) No 1744/2006 of 24 Nov. 2006 on detailed rules for aid in respect of silkworms.
142. The Commission's competence is limited to olive oil and does not apply to table olives; see Art. 113(1)(a)'s reference to point (a) of Part VII of Annex I.

113(2)(a) sets certain limits to the Commission's competence, but it is clear from subparagraph (b) that the Commission has extensive competence. The term 'marketing standards' may relate in particular to quality, grading, weight, sizing, packaging, wrapping, storage, transport, presentation, origin and labelling. According to Article 113(3), unless the Commission determines otherwise, the products for which marketing standards have been laid down may only be marketed in the Union in accordance with such standards.

Article 113(3) states that it is up to the Member States to check whether products conform to the standards laid down and to apply penalties as appropriate. However, the Member States' sanctioning of breaches of standards is subject to the general powers of the Commission under Article 194 to adopt administrative penalties for breaches of the rules of the Single CMO Regulation.

In addition to the general standards laid down by the Commission pursuant to Article 113, further standards for specific categories of goods are laid down in Articles 113a, 113b, 113c and 114–118.

[2] Marketing Standards for Fruit and Vegetables

Article 113a(1) provides that the products of the fruit and vegetables sector which are intended to be sold fresh to the consumer may only be marketed if they are sound, fair and of marketable quality and if the country of origin is indicated. Article 113a(2) stipulates that these marketing standards apply at all marketing stages including import and export, unless otherwise provided by the Commission.

[3] Marketing Standards for Young Cattle

Article 113b lays down the main rule that the conditions in Annex Xia of the Single CMO Regulation apply to the meat of bovine animals aged 12 months or less. Meat from animals that are a maximum of 8 months is designated as 'veal', and meat from animals over 8 months but below 12 months is designated 'beef'.

[4] Marketing Standards for Wine

As part of the incorporation of the market organization for wine in the Single CMO Regulation, Article 113c and 113d were included in the chapter on marketing rules. Article 113c allows individual Member States to lay down marketing rules to regulate the supply of wine. The main purpose of these marketing rules is to improve and stabilize the operation of the common market in wines, including the grapes, musts and wines from which they derive. It is expressly stated in Article 113c(1) that such rules shall be proportionate to the objective pursued.[143] The provision also states what national marketing rules must not concern. First, such rules may not relate to any

---

143. See Ch. 3, §3.02[C], on the proportionality principle.

transaction after the first marketing of the produce concerned.[144] Next, the rules must not allow for price-fixing, including the setting of recommended prices.[145] As a third restriction on the Member States' freedom of action, the marketing rules may not make unavailable an excessive proportion of the vintage that would otherwise be available. This restriction is very vague. The interpretation of 'an excessive proportion' must primarily be made in the light of the overall purpose of Article 113c, so that the national marketing rule may not damage or destabilize the operation of the market. Finally, national rules may not allow refusal to issue the certificates required for the circulation and marketing of wines where such marketing is in accordance with the national marketing rules.

Under Article 113d(1), the designations for categories of grapevine products, as provided in Annex XIb to the Single CMO Regulation, may only be used in the Union for the marketing of a product which conforms to the corresponding conditions laid down in the Annex. The Annex lists 17 different categories of grapevine products and for each of them states its special characteristics and the designation under which it may be marketed. Article 113d(2) authorizes the Commission to amend the categories of grapevine products listed in the Annex. In this case the Commission must follow the procedure set out in Article 195(4), which in turn refers to the examination procedure which is described in greater detail in Article 5 of Regulation (EU) No 182/2011.[146] Notwithstanding the rules on the use of designations for grapevine products, according to Article 113d(1) there are two situations in which the Member States may permit the use of the term 'wine' other than in accordance with the categories listed in Annex XIb. This term can be used if it is accompanied by the name of a fruit in the form of a composite name in order to market products obtained by the fermentation of fruit other than grapes. The word 'wine' can also be used if it is part of a composite name. It is emphasized that any confusion with products corresponding to the wine categories in Annex XIb must be avoided.

[5]   Marketing Standards for Milk and Milk Products

According to Article 114, foodstuffs intended for human consumption may only be marketed as milk and milk products if they comply with the definitions and designations laid down in Annex XII and Annex XIII. Point I of Annex XII defines 'marketing' as holding or display with a view to sale, offering for sale, sale, delivery or any other manner of placing on the market. In Annex XIII, 'milk' is defined as the produce of the milking of one or more cows, and 'drinking milk' is defined as the products referred to

---

144. On the interpretation of 'undergo first-stage processing' in connection with agricultural products, see Ch. 2, §2.02.
145. On the competition rules in the agricultural sector, see Ch. 9.
146. Article 195(4) of the Single CMO Regulation refers to Arts 5 and 7 in Council Decision 1999/468/EC laying down the procedures for the exercise of implementing powers conferred on the Commission. This decision was repealed by Regulation (EU) No 182/2011, and it is stated in Art. 13(1)(c) of this Regulation that for references in legislation to Art. 5 of the Decision, Art. 5 of Regulation (EU) No 182/2011 applies. On the examination procedure, see Ch. 3, §3.05[C].

in point III. According to this, 'drinking milk' includes raw milk, whole milk, semi-skimmed milk and skimmed milk. The individual categories are defined in more detail in the Annex. According to Point II, paragraph 1 of Annex XII, the term 'milk' means exclusively the normal mammary secretion obtained from one or more milkings without either addition thereto or extraction therefrom.[147] Under Point II, paragraph 2 of Annex XII, the term 'milk products' means products derived exclusively from milk. However, substances necessary for their manufacture may be added, provided those substances are not used to replace, in whole or in part, any milk constituent.

[6]   Marketing Standards for Fats

Marketing standards for certain groups of fats are laid down in Article 115 of the Single CMO Regulation, supplemented by Annex XV. The Appendix to the Annex contains detailed rules as to which designations can be used for specific kinds of fats.[148]

[7]   Marketing Standards for Egg and Poultry Meat Sector

Marketing standards for the products of the eggs and poultry meat sectors are laid down in Article 116 of the Single CMO Regulation, which refers to Annex XIV of the Regulation. The Annex contains standards for three sectors in the form of marketing standards for eggs of hens of the *Gallus gallus* species, marketing standards for poultry meat, and marketing standards for the production and marketing of eggs for hatching and farmyard poultry chicks. The marketing standards for eggs and poultry meat include rules on quality grading for individual products, and rules on the labelling of the products with information about their quality grading.

[8]   Certification for Hops

Article 117 contains rules on the certification of hops. According to this, products of the hops sector harvested or prepared within the Union are subject to a certification procedure. Article 117(4) states that products of the hops sector may only be marketed or exported if a certificate has been issued in respect of them. Under Article 117(2) and (3), certificates may only be issued for products that have minimum quality characteristics appropriate to a specific stage of marketing. The certificates must at least indicate the place(s) of production of the hops, the year(s) of harvesting, and the variety or varieties. The Commission has issued a Regulation with implementing provisions for the certification of hops and hop products.[149]

---

147. There are a few exceptions to the rule on the use of the term 'milk'; see Point II, paragraphs (a) and (b), in Annex XII.
148. For the use of the sales designations 'butter' and 'dairy spread', see Case C-37/11 *Commission v. Czèch Republic* [2012] ECR.
149. Commission Regulation (EC) No 1850/2006 laying down detailed rules for the certification of hops and hop products. Following the entry into force of the Single CMO Regulation, there is

[9]   Marketing Standards for Olive Oils and Olive-Pomace Oils

Marketing standards for olive oils and olive-pomace oils are laid down in Article 118. According to paragraph (1) of the Article, the use of the descriptions and definitions of olive oils and olive-pomace oils set out in Annex XVI is compulsory. This applies both to the marketing of the products concerned within the Union and sales to third countries. However, in relation to trade with third countries, this only applies insofar as it is compatible with compulsory international rules. Annex XVI contains designations and definitions for six different product types. According to Article 118(2), of these categories and their sub-categories, only extra virgin olive oil, virgin olive oil, olive oil composed of refined olive oils and virgin olive oils, and olive-pomace oil may be marketed at the retail stage.[150] Among other things under the authority of Article 113(1)(a) and Article 121, first paragraph, point (a), in conjunction with Article 4 of the Single CMO Regulation, the Commission has adopted Commission Implementing Regulation (EU) No 357/2012 on marketing standards for olive oil

[10]   Designations of Origin, Geographical Indications and Traditional Terms (Wine)

The inclusion of the wine sector under the rules of the Single CMO Regulation on marketing and production did not only involve the insertion of rules on marketing standards for wine. In Article 118a–118zb there are special rules on designations of origin, geographical indications and traditional terms in the wine sector. These rules apply to some of the grapevine products listed in Annex XIb. According to Article 118a(1), these rules apply to: wine, liqueur wine, sparkling wine, quality sparkling wine, quality aromatic sparkling wine, semi-sparkling wine, aerated semi-sparkling wine, partially fermented grape must, wine from raisined grapes, and wine from overripe grapes. Article 118a(2) sets out the basic principles for these rules. The rules are primarily based on protecting of legitimate interests of consumers and producers. These aims should presumably be interpreted in the light of Article 39(1)(e) and (b) TFEU, and the general Union law principle of the protection of legitimate expectations.[151]

The terms 'designation of origin' and 'geographical indication' are defined in Article 118b(1). In the main, both terms are defined in the same way, but they differ in respect of the special conditions that are additionally attached to the terms. According to Article 118b(1)(a) and (b), both designations of origin and geographical indications mean the name of (or an indication referring to) a region, a specific place or, in exceptional cases, a country used to describe a product referred to in Article 118a(1).

---

authority in Art. 121 of the Single CMO Regulation for the Commission to issue implementing regulations for marketing standards for hops.
150.   Article 118(2) refers to points 1(a) and (b), 3 and 6 of Annex XVI.
151.   On the interpretation of Art. 39(1)(e) and (b) TFEU, see Ch. 2, §2.05[B][5] and §2.05[B][2]; and on the Union law principle of the protection of legitimate expectations, see Ch. 3, §3.02[D].

Article 118b(1)(a) states that a designation of origin indicates that, as special conditions, a product's quality and characteristics are essentially or exclusively due to a particular geographical environment with its inherent natural and human factors; that the grapes from which it is produced come exclusively from this geographical area; that its production takes place in this geographical area; and that it is made from grape varieties belonging to the *vitis vinifera* species.

Certain traditionally used names may constitute a designation of origin where they designate a wine, refer to a geographical name and otherwise fulfil the requirements for designations of origin in Article 118b(1), and undergo the procedure conferring protection on designations of origin and geographical indications laid down in Article 118c–118t.

As stated, a geographical indication is defined in the same way as a designation of origin. As special conditions, Article 118b(1)(b) states that a product must possesses a specific quality, reputation or other characteristics attributable to that geographical origin; at least 85% of the grapes used for its production must come exclusively from that geographical area; its production must take place in the geographical area; and it must be obtained from *vitis vinifera* grape varieties or a cross between *vitis vinifera* and other species of the genus *vitis*.

Article 118k(1) provides that the name of a vine product that has become generic may not be protected as a designation of origin or geographical indication. A name for a wine is considered as having become generic if it has become the common name of a wine in the Union, even though it relates to the place or the region where the product was originally produced or marketed. The provision states that, in order to establish whether a name has become generic, account shall be taken of the existing situation in the Union, especially in the areas of its consumption, and the relevant Union or national legislation.[152]

The procedural rules in Article 118c–118t, which must be complied with in order to obtain the protection of a designation of origin or geographical indication, are highly detailed. In principle, only producer organizations can apply for protection of a designation of origin or geographical indication. According to Article 118e(1), only exceptionally can a single producer apply for protection. Single producers may only lodge an application for protection for wines which they themselves produce.

Under Article 118f, an application for protection of a designation of origin or a geographical indication must be subject to a preliminary national procedure in the Member State where the designation of origin or geographical indication originates.[153] If the Member State's authorities find that the designation of origin or a geographical indication does not meet the relevant requirements or is incompatible with Union law in general, then according to Article 118f(4) they must reject the application. There is no express provision in the procedural rules giving an unsuccessful applicant a right to appeal to the Commission against a rejection by a national authority. If a Member State finds that the relevant requirements have been met, then among other things it must

---

152. A protected traditional term cannot become generic in the Union; see Art. 118v(2).
153. In Art. 118d there is a special provision for applications for protection for a geographical area in a third country.

forward the application to the Commission which then examines whether the conditions for the grant of protection have been fulfilled. In Article 118f(5) and 118g, there
are rules on the publication of the decisions of authorities on protection, and in Article
118n there is a provision on the establishment and maintenance of an electronic
register of protected designations of origin and geographical indications for wine which
must be publicly accessible.

Article 118u–118y contain rules on the protection of 'traditional terms'.[154]
According to Article 118u(1), a 'traditional term' means a term traditionally used in
Member States for products referred to in Article 118a(1). The term is used to indicate
that the product has a protected designation of origin or a protected geographical
indication under Union or national law. Alternatively, the traditional term can be used
to indicate the production or ageing method or the quality, colour, type of place, or a
particular event linked to the history of the product with a protected designation of
origin or a protected geographical indication.

[11]   Special Rules on Labelling and Presentation in the Wine Sector

The special rules on labelling and presentation in the wine sector are set out in Article
118w–118zb. According to Article 118x, 'labelling' means any words, particulars,
trademarks, brand name, pictorial matter or symbol placed on any packaging, document, notice, label, ring or collar accompanying or referring to a given product.
According to the same provision, 'presentation' means any information conveyed to
consumers by virtue of the packaging of the product concerned, including the form and
type of bottles. The fact that the Single CMO Regulation contains special rules on
labelling and presentation in the wine sector does not mean that the general rules of
Union law do not apply to this sector. Article 118x expressly states that the general
rules do apply in the wine sector unless otherwise indicated by the Single CMO
Regulation.[155]

In principle, the rule on obligatory labelling and presentation of grapevine
products applies to all products referred to in Annex XIb. Article 118y states that the
labelling and presentation of grapevine products must include a number of obligatory
particulars.[156] Commission Regulation (EC) No 607/2009 lays down detailed implementing provisions for protected designations of origin and protected geographical
indications and the labelling and presentation of certain grapevine products. According

---

154. The Commission's original proposal for a market organization for wine (COM(2007) 372 final),
     which resulted in Council Regulation (EC) No 479/2008, did not contain independent rules on
     the protection of traditional terms.
155. Article 118x refers to the general rules on labelling and presentation of foods in Directive
     89/104/EEC, Council Directive 89/396/EEC, Directive 2000/13/EC, and Directive 2007/45/EC.
     Directive 89/104/EEC has now been repealed and replaced by Directive 2008/95/EC, and
     Directive 89/396/EEC has now been repealed and replaced by Directive 2011/91/EU.
156. Of the group of grapevine products listed in Annex XIb to the Single CMO Regulation which are
     subject to obligatory rules on labelling and presentation, Art. 118y(1) excludes only partially
     fermented grape must obtained from raisined grapes, rectified concentrated grape must and
     wine vinegar. See §4.05[A][10] above on the different categories of grapevine products listed in
     Annex XIb.

to Article 118y(1)(a), the labelling must state the designation for the category of the grapevine product. However, this rule can be derogated from pursuant to Article 118b(2) for wines whose labels include the name of a protected designation of origin or a protected geographical indication. The second obligatory requirement, under Article 118y(2), is that wine with a protected designation of origin or a protected geographical indication must be labelled stating that it has a protected designation of origin or protected geographical indication; and must state the protected designation of origin or protected geographical indication. This rule too may be derogated from under Article 118y(3) if a traditional term, as referred to in Article 118u(1)(a), is displayed on the label or if the protected designation of origin or protected geographical indication is displayed on the label.[157] Third, pursuant to Article 118y(1)(c), the label must state the actual alcoholic strength by volume. Article 118y(1)(d) requires the label to give an indication of provenance. There can be some doubt as to what precisely is meant by 'provenance'. In its implementing Regulation (EC) No 607/2009, the Commission interpreted the term as meaning that an indication of provenance was in the interests of product traceability and transparency.[158] Under Article 55 of Regulation (EC) No 607/2009, the indication of provenance must give the name of the Member State or third country where the grapes were harvested and processed into the grapevine product in question. As the fifth requirement for obligatory labelling, Article 118y(1)(e) requires there to be an indication of the bottler. However, in the case of sparkling wine, aerated sparkling wine, quality sparkling wine or quality aromatic sparkling wine, the name of the producer or vendor must be given. As for imported wine, Article 118y(1)(f) states that the name of the importer must be given. Under Article 118y(1)(g), for sparkling wine, aerated sparkling wine, quality sparkling wine or quality aromatic sparkling wine, an indication of the sugar content must be given on the label.

In addition to the rules on the obligatory labelling of grapevine products, Article 118z contains rules on optional particulars. These statements may be about the vintage year or about the name of one or more wine grape varieties.

Whether a statement is obligatory or optional, Article 118za states that where a statement is expressed in words, it must be in one or more of the official languages of the Union. However, this does not apply to the naming of a protected designation of origin or a protected geographical indication or a traditional term. In this case the name must appear on the label in the language or languages for which the protection applies. If the designation in question uses a non-Latin alphabet, it can also be given in one or more of the official languages of the Union.

The rules in Article 118y, 118z and 118za must be taken seriously, as their enforcement is backed up by Article 118zb. According to this provision, the competent authorities of the Member States must take measures to ensure that a product referred to in Article 118y(1) which is not labelled in conformity with the rules referred to above is not placed on, or is withdrawn from, the market.

---

157. The possibility of merely stating the protected designation on the label only applies under 'exceptional circumstances' which, according to Art. 118y(3), are to be determined by the Commission.
158. See recital 22 of Commission Regulation (EC) No 607/2009.

*[12]   Special Rules on the Conditions for Production of Certain
        Agricultural Products*

In addition to the detailed rules on the sale of agricultural products, the Single CMO Regulation also contains a few rules for how certain products must be produced. These rules concern the use of casein and caseinates in the manufacture of cheese, methods of production of agricultural ethyl alcohol, and grapevine products.

Where aid is given under Article 100 of the Single CMO Regulation on aid for skimmed milk processed into casein and caseinates, under Article 199 the Commission may make the use of casein and caseinates in the manufacture of cheese subject to prior authorization. Such authorization can only be given if such use is a necessary condition for the manufacture of the products. Under Article 121(i), it is the Member States that grant the authorizations, but the Commission which lays down the conditions for the grant of such authorizations.[159]

Article 120 authorizes the Commission to lay down rules on the method of production and the characteristics of agricultural ethyl alcohol obtained from a specific agricultural product listed in Annex I to the TFEU.[160]

The third and last group of rules on the production of a particular category of agricultural products is in Article 120a–120g on the wine sector. This group of rules is the most extensive and the most detailed. Article 120a contains provisions on the classification of wine grape varieties. According to its paragraph (1), the products listed in Annex XIb to the Single CMO Regulation and produced in the Union must be made from wine grape varieties classifiable according to paragraph (2).[161] Under Article 120a(2), it is the Member States that classify which wine grape varieties may be planted, replanted or grafted on their territories for the purpose of wine production. However, the powers of the Member States in this respect are not unlimited. Only wine grape varieties of the *Vitis vinifera* kind or of a cross between the species *Vitis vinifera* and other species of the genus *Vitis* may be classified by Member States, not including the varieties Noah, Othello, Isabelle, Jacquez, Clinton or Herbemont. If a Member State deletes a wine grape variety from its classification, it must be grubbed up within 15 years of its deletion. There is a general provision in Article 120a(5), whereby areas planted with wine grape varieties that have been dropped from the classification must

---

159. See Commission Regulation (EC) No 548/2009 as regards authorizations for the use of casein and caseinates in the manufacture of cheeses.
160. Article 201(1)(a) of the Single CMO Regulation repealed Council Regulation (EC) No 670/2003 laying down specific measures concerning the market in ethyl alcohol of agricultural origin; under the authority of this Regulation the Commission had issued Commission Regulation (EC) No 2336/2003 introducing certain detailed rules for applying Council Regulation (EC) No 670/2003 laying down specific measures concerning the market in ethyl alcohol of agricultural origin. Commission Regulation (EC) No 2336/2003 is still in force under the authority of Art. 201(3) of the Single CMO Regulation. On the meaning of 'agricultural product' see Ch. 2, §2.02.
161. A Member State is exempt from the obligation regarding wine grape varieties if its wine production does not exceed 50,000 hectolitres per wine year, calculated on the basis of the average production during the latest five wine years; see Art. 120a(3).

be grubbed up. However, there is no obligation to grub-up areas whose production is intended exclusively for consumption by the wine-producer's household.[162]

Oenological processes play an important role in wine production. In Article 120b–120g, there are detailed rules on permitted oenological practices and restrictions. In connection with this there are rules about the procedures which must be followed in connection with authorizing the use of different practices. According to Article 120f, it is the Commission which, within certain limits, grants authorization. However, under Article 120d, Member States may limit or exclude the use of certain oenological practices and provide for more stringent restrictions for wines than those authorized under Union law. However, this only applies to wine produced in their territory, and only with a view to reinforcing the preservation of the essential characteristics of wines with a protected designation of origin or a protected geographical indication and of sparkling wines and liqueur wines.

[13] *Authority for the Commission to Adopt Implementing Provisions*

Article 121 gives the Commission the power to adopt implementing provisions for a number of the sectors and products that are dealt with in the articles referred to in sections §4.05[A][4]–§4.05[A][10] above; i.e., Articles 113–120 and their associated annexes.[163]

[B] Producer Organizations and Groups as Part of the Market Organization

Article 122 contains provision on producer organizations. According to Article 122(1), Member States must recognize producer organizations which are constituted by producers of one of the following sectors: the hops sector, the olive oil and table olives sector,[164] the fruit and vegetables sector,[165] and the silkworm sector.[166] The obligation to recognize only applies to organizations that are formed on the initiative of the producers and which pursue a specific aim. This specific aim may, and in the case of the fruit and vegetables sector it must, relate to one or more specified aims. First, the producer organization may have the aim ensuring that production is planned and adjusted to match demand, particularly with regard to quality and quantity. Next, the organization may have the aim of concentrating supply and the placing on the market of the products produced by its members. Finally, the aim can be the optimization of production costs and stabilization of producer prices.

There are two exceptions to the above-named principle of the Member States' recognition of producer organizations in certain sectors. In addition to producer

---

162. On the grubbing-up of vine-growing areas in connection with market intervention in the form of production regulation, see above in §4.04[C][4].
163. On the general rules on the delegation of legislative powers and implementing authority to the Commission, see Ch. 3, §3.05[C].
164. On aid for the olive oil and table olives sector, see above in §4.04[D][5].
165. On aid for the fruit and vegetables sector, see above in §4.04[D][6].
166. On aid for the silkworm sector, see above in §4.04[D][11].

organizations in the sectors referred to, which Member States must recognize, under Article 122, second paragraph, Member States may recognize producer organizations that are set up by producers in the sectors referred to in Article 1 of the Single CMO Regulation.[167] The other exception pursuant to Article 122, third paragraph, is for producer organizations in the wine sector. In recognizing producer organizations pursuant to Article 122, second and third paragraphs, the condition also applies that the organization must have been set up on the initiative of the producers and set up to pursue a specific aim. Article 122, third paragraph, contains detailed rules for the wine sector. A producer organization in this sector that wishes to be recognized by a Member State must apply rules of association which impose three kinds of obligations on its members. First, its members must apply the rules adopted by the producer organization relating to production reporting, production, marketing and protection of the environment. Second, the organization's rules must require the members to provide the information requested by the producer organization for statistical purposes, in particular on growing areas and market evolution. And third, the members must be bound to pay penalties for the infringement of obligations under the rules of association. Article 122, third paragraph, also states the specific aims which individual producer organizations in the wine sector must pursue. The provisions must be interpreted as meaning that these aims apply in addition to the general aims which a producer organization must have pursuant to Article 122, first paragraph. Article 122, third paragraph, provides that producer organizations in the wine sector must seek to promote and provide technical assistance for the use of environmentally sound cultivation practices and production techniques. Next, the individual organization must have the aim of promoting initiatives for the management of by-products of wine-making and the management of waste in particular to protect the quality of water, soil and landscape and preserving or encouraging biodiversity. Third, producer organizations can have the aim of carrying out research into sustainable production methods and market developments. And finally, such organizations can have the aim of contributing to the achievement of support programmes as referred to in the rules in Article 103i–103z on support schemes for the wine sector.[168]

Article 123 sets out the general rules for inter-branch organizations. The rules govern such organizations within the sectors for olive oil and table olives, tobacco, fruit and vegetables, and wine. Under certain conditions, the Member States must recognize inter-branch organizations in these sectors. Recognition requires that the organization must have been formed on the initiative of all or some of the organizations or associations which constitute them. In addition, inter-branch organizations must pursue a specific aim, which may, 'in particular', relate to the four aims listed in Article 123(1)(c). The use of the words 'in particular' must mean that aims other than the four aims listed can also be sufficient. The four aims listed are: concentrating and coordinating supply and marketing of the produce of the members; adapting production and processing jointly to the requirements of the market and improving the product;

---

167. On the scope of the Single CMO Regulation, see §4.02[B] above.
168. See §4.04[D][8] above, on market intervention in the form of support schemes in the wine sector.

promoting the rationalization and improvement of production and processing; and carrying out research into sustainable production methods and market developments. Article 123(3) states that, further to paragraph (1) and subject to certain conditions, Member States must, with regard to the fruit and vegetables sector, and may, with regard to the wine sector, also recognize inter-branch organizations.[169]

Article 124(1) lays down some common provisions concerning producer and inter-branch organizations. Article 124 is a new provision which did not appear in any of the earlier market organizations. According to the provision, Article 122 and Article 123(1) apply in the sectors referred to in Article 1 of the Single CMO Regulation, but without affecting recognitions of producer or inter-branch organizations already decided by Member States on the basis of national law. However, this only applies if the national law is in compliance with Union law, and in any case this does not apply to the sectors referred to in Articles 122 and 123(1).

In certain places, the Single CMO Regulation uses the term 'operator organisations'. According to Article 125, this term means recognized producer organizations, recognized inter-branch organizations or recognized organizations of other operators in the olive oil and table olives sector or their associations.[170]

In the fruit and vegetables sector, the problem has arisen that some producers have not wanted to be members of the relevant producer or inter-branch organization. The consequence of this lack of membership has been that producers have not been obliged to follow the organizations' decisions about quality standards, grading, sale through auctioneers approved by the organization and withdrawal from the market.

In a leading case, which concerned a previous market organization for fruit and vegetables, a French farmer, Albert Le Campion, refused to be a member of a producer organization and refused to comply with the producer organization's rules and decisions. According to a provision in the French law on agriculture, a producer organization's rules and decisions could be extended so as to apply to all producers in the geographical area covered by the organization. However, Le Campion denied that he could be bound by the rules of the local producer organization to report the size of his cultivated areas and that he should pay a members' contribution. Le Campion argued before the French courts that, among other things, the French legislation, which extended the rules and decisions of a private producer organization so as to apply to non-members, was contrary to the provisions of the Regulation establishing the common market organization for fruit and vegetables. The case was referred to the Court of Justice which pointed out that where a Union regulation has established a common market organization that is intended to establish an exhaustive scheme of

---

169. Fulfilment of the requirements of Art. 123 is linked among other things to the inter-branch organization's possibility of claiming exemption from the competition rules; see Arts 176a to 178, and see below in Ch. 8, §8.03[B]. Arts 125a to 125n, 125o and 126 give special and highly detailed rules for producer and inter-branch organizations and operator organizations in the fruit and vegetables, wine and tobacco sectors.
170. On operator organizations see §4.04[D][5] on Art. 103 on aid for the olive oil and table olives sector aid for three-year work programmes drawn up by operator organizations, and §4.04[D][10] on support for apiculture programmes drawn up by representative organizations and beekeeping cooperatives pursuant to Art. 109.

regulation, the Member States no longer have any power to add supplementary regulations. The rules on quality standards in the common organization for fruit and vegetables were of an exhaustive nature, and it was therefore in breach of these rules for the rules of the producer organization on grading, size, weight and presentation to be given binding effect on producers who were not members of the organization. The Court of Justice ruled that such an extension of the rules was not authorized by the Union provision which governed the area.[171]

Likewise, the Court of Justice said that there was a very clear distinction between the intervention mechanisms which may be initiated by producers' groups and those which are applicable to all producers. Consequently, a Member State has no power to extend the intervention rules laid down by producers' organizations so as to apply to all producers.[172]

As for the national provisions which required producers to report their areas of cultivation to the producer organization, the Court of Justice stated that since the common market organization did not contain any rules on this, the national provisions were not contrary to Union law. With regard to the obligation to pay a contribution to producer organizations, the Court of Justice stated that the lawfulness of the national rules depended on whether the obligation could be regarded as financing activities which were contrary to Union law. This was a matter for the national court to decide.[173]

Article 125f and 125i of the Single CMO Regulation contain express rules on the extent to which non-members of producer organizations in the fruit and vegetables sector can be required to contribute to a producer organization and be bound by the organization's rules.[174]

Article 125o contains special provisions on producer and inter-branch organizations in the wine sector. Article 126 contains a special rule for inter-branch organizations in the tobacco sector, on the question of non-members paying contributions to the organization.

Under Article 127, the Commission has authority to adopt detailed rules for this area, in particular the conditions and procedures for the recognition of producer, inter-branch and operator organizations.

§4.06   TRADE WITH THIRD COUNTRIES

Part III of the Single CMO Regulation is headed 'Trade with third countries', and it is divided into two main parts, one chapter on imports and one chapter on exports. The rules are based on the obligations which the European Community, now the European Union, undertook in connection with the Uruguay Round of the General Agreement on

---

171. Case 218/85 *Le Campion* [1986] ECR 3513, paras 12–16.
172. *Ibid.*, paras 17–19.
173. *Ibid.*, paras 21–22. In Case 222/82 *Apple and Pear Development Council* [1983] ECR 4083, the Court of Justice had ruled that national laws, whereby a producer was obliged to be a member of a producer organization whose aims were product development and the marketing of products, could only be regarded as being in breach of the concrete market organization to the extent that the activities of the organization were in themselves in breach of the rules of the market organization.
174. See also Arts 125l and 125n on inter-branch organizations.

Tariffs and Trade (GATT). This is clearly expressed in Article 128 which lays down the general principle that, unless otherwise provided for in the Regulation (or in provisions adopted pursuant to it), the levying of any charge having equivalent effect to a customs duty or the application of any quantitative restriction or measure having equivalent effect is prohibited in trade with third countries. Article 128 implements in Union law Article 4 of the WTO Agreement on Agriculture.[175]

The tariffs on the products covered by the Single CMO Regulation are very important in connection with trade with third countries. According to Article 129, the general rules for interpreting the Combined Nomenclature and the special rules for its application apply to the tariff classification of products covered by the Single CMO Regulation.

[A] Imports from Third Countries

[1] Import Licenses

Before the Single CMO Regulation entered into force, control of the extent of trade in agricultural products with third countries took place within the frameworks of a large number of different market organizations. The controls often took the form of obligatory licensing arrangements or arrangements under which the Commission had powers to lay down licensing requirements. On the basis of the experiences with the different market organizations in which the Commission already had responsibility for administering the licensing arrangements, in the Single CMO Regulation the Union legislator has emphasized that the control of trade is primarily an administrative matter which should be dealt with flexibly.[176] As a consequence, the Single CMO Regulation introduces an arrangement whereby the Commission decides on the requirement for a license within an individual sector.[177] Article 130(1) provides that, in addition to the situations where import licenses are required in accordance with the Single CMO Regulation, the Commission may make imports into the Union of products in a number of sectors subject to possession of an import license. The Single CMO Regulation requires import licenses in the sugar sector.[178] In connection with decisions to introduce a license requirement in a given sector, the Commission must have regard for whether there is a need for import licenses for the management of the market in question and in particular whether there is a need to monitor imports of the products in question.[179]

The Commission's powers with regard to import licenses apply to virtually all products and all sectors that are covered by the Single CMO Regulation. The sectors

---

175. On the Agreement on Agriculture and the Union's international obligations, see Ch. 9.
176. On the law and the principal problems associated with the administration of export licenses, see O'Connor, *Agricultural Law for the European Union* 295 (Heusel & Collins eds., Academy of European Law and Irish Centre for European Law, 1999).
177. Recitals 66 and 67 of the Single CMO Regulation.
178. See Art. 153(3) on import licenses for sugar for refining, and Art. 153 on import licenses for sugar that is covered by a guaranteed price arrangement.
179. Recital 67, third sentence, and Art. 130(2) of the Single CMO Regulation.

covered are listed in Article 130(1), and a comparison with Article 1 shows that only dried fodder, raw tobacco, apiculture products, silkworms and potatoes are excluded.[180]

Even though it is the Commission that has powers to introduce and lay down rules for licensing arrangements, it is the Member States that issue import licenses. The authorities of a Member State may not restrict themselves to issuing licenses to their own citizens or to undertakings based in their own territory. The principle is that the proper national authorities must issue import licenses to any applicant, regardless of where in the Union they are based. Under Article 131(1), this applies unless the Council provides otherwise, and according to Article 132 import licenses must be valid throughout the Union.

Article 133(1) of the Single CMO Regulation established the basic rule that, unless otherwise provided for by the Commission, licenses are issued subject to the lodging of a security guaranteeing that the products will be imported during the term of validity of the license.[181] If security is lodged, i.e., if the Commission does not derogate from the rule in Article 133(1), Article 133(2) provides that the security will be forfeited in whole or in part if the import is not carried out, or is carried out only partially, within the period of validity of the license. Naturally enough, the Commission does not have powers to decide that the security shall not be lost in the event of the import not being carried out within the period of validity of the license. Article 133(2) carries forward a rule from earlier market organizations that, even if an import is not carried out, the security will not be lost in the event of force majeure.[182]

According to Article 134, it is the Commission that adopts the detailed implementing rules on import licenses, including the terms of validity of the licenses and the rate of security. The Commission has exercised its power under Article 134 by adopting Commission Regulation (EC) No 376/2008 laying down common detailed rules for the application of the system of import and export licenses and advance fixing certificates for agricultural products.[183] Article 7(1) of Regulation (EC) No 376/2008 expresses the

---

180. In addition, certain 'other products', such as live horses, are excluded from the Commission's licensing powers. Article 153(1) to (3) contains special rules on import licenses for sugar for refining. There are special rules on, among other things, quality requirements for the import of hemp and hops in Arts 157 and 158.
181. There is a provision on special security in the wine sector in Art. 133a.
182. See §4.06[C] below on force majeure and the loss of security lodged in connection with licenses etc.
183. See also Commission Implementing Regulation (EU) No 282/2012 laying down common detailed rules for the application of the system of securities for agricultural products. The system requiring undertakings to lodge security for the use of import and export licenses has been at issue in a number of cases. See Case 11-70 *Internationale Handelsgesellschaft* [1970] ECR 1125, on the system and its compatibility with the proportionality principle (on this principle, see Ch. 3, §3.02[C]); Case 137/85 *Maizena* [1987] ECR 4587 (on whether loss of security constitutes a sanction); Case 26/70 *Einfuhr- und Vorratsstelle für Getreide und Futtermittel* [1970] ECR 1183 (on the amount of security required); and Case 186/73 *Norddeutsches Vieh- und Fleischkontor* [1974] ECR 533 (on fulfilment of the conditions for release of the security). On the provision of security under Union law in agricultural cases, see Barents, 'The System of Deposits in Community Agricultural Law: Efficiency v. Proportionality', *ELR* 1985, s. 239.

core of the licensing system whereby an import or export license constitutes authorization and gives rise to an obligation to import or to export under the license the specified quantity of the goods concerned during its period of validity, except in cases of force majeure.

[2] Import Duties and Additional Import Duties

The starting point of the Single CMO Regulation is that imports of the agricultural products referred to in Article 1 of the Regulation are covered by the import duties covered by the Common Customs Tariff. This starting point, which is stated in Article 135, is derogated from in Articles 136–140a for a number of products (certain kinds of cereals, certain kinds of rice, and fruit and vegetables and processed fruit and vegetables), and the tariff can be changed for individual products. Such derogation is made either under fixed rules in the Single CMO Regulation, as in the case of cereals, or it is left to the decision of the Commission, as in the case of rice and fruit and vegetables.[184] As special rules for sugar and isoglucose, Article 142 provides that the Commission may suspend import duties in whole or in part in order to guarantee the supply necessary for the manufacture of products referred to in Article 62(2).[185] Where an imported consignment consists of mixtures of different kinds of cereals or different kinds of rice, Articles 149–152 contain rules on import duties and the tariff applicable to the goods.

In addition to the standard import duty, under Articles 135–140a, an additional import duty can be imposed on certain products. Article 141 allows for an additional import duty to be applied to products of the cereals, rice, sugar, beef and veal, milk and milk products, pig meat, sheep meat and goat meat, eggs, poultry meat and bananas sectors. According to Article 143(b), it is the Commission that adopts the necessary implementing rules for import duties and levies, and therefore also decides on the imposition of additional import duties.[186] According to Article 141(1), such an additional import duty can only be used to prevent or counteract adverse effects on the market of the Union which may result from those imports, and according to paragraph (2), such additional duties cannot be imposed if the imports are unlikely to disturb the Union market, or if the effects would be reasonable in relation to their aim.[187] The imposition of additional duty is also made conditional on the existence of one of two alternatives. First, under paragraph (1)(a), the import of the product in question must be made at a price below the level notified by the Union to the WTO (the trigger price). Article 141(3) provides that import prices are to be determined on the basis of the c.i.f. (cost, insurance, freight) import prices of the consignment under consideration. The

---

184. For fruit and vegetables there is a special entry price arrangement; see Art. 140a.
185. Article 153(4) contains a special rule on the suspension of import duty on cane sugar for refining from certain third countries.
186. See Commission Regulation (EC) No 1484/95 laying down detailed rules for implementing the system of additional import duties and fixing additional import duties in the poultry meat and egg sectors and for egg albumin.
187. Article 141(2) uses the wording: 'where the effects would be disproportionate to the intended objective'. On the proportionality principle, see Ch. 3, §3.02[C].

actual c.i.f. import prices must be checked against the representative prices for the product on the world market or on the Union import market. Second, under paragraph (1)(b), imposition of additional duty is permissible if the volume of imports in any year exceeds a certain level (the trigger volume). The trigger volume is based on market access opportunities defined, where applicable, as imports as a percentage of the corresponding domestic consumption during the three previous years.

[3]   Tariff Quotas in Connection with Imports

Articles 144–148 contain special rules on tariff quotas in connection with the import of agricultural products.[188] A tariff quota determines in advance a fixed value or a fixed quantity of a specified product that can be imported in a given period at a reduced tariff. When the quota has been used up, the product in question can still be imported, but duty is payable at the normal rate. Tariff quotas can arise in agreement with third countries entered into by the Council under the authority of Article 218 TFEU, or under other legislative acts adopted by the Council. According to Articles 144(1) and 145 of the Single CMO Regulation, tariff quotas are opened and administered by the Commission pursuant to implementing provisions which are also laid down by the Commission. Article 148 states that the detailed implementing rules must, 'in particular', include guarantees covering the nature, provenance and origin of the product.

The allocation of tariff quotas between economic operators causes particular problems and this is governed by Article 144(2). It is a fundamental requirement for the Commission's administration of tariff quotas that they should be administered in a manner which avoids any discrimination between the operators concerned.[189] At the same time, under Article 144(3), in choosing its method of administration the Commission must, 'where appropriate', give due weight to the supply requirements of the Union market and the need to safeguard the equilibrium of that market. Subject to complying with these requirements, the Commission is free to allocate quotas among the economic operators. However, Article 144(2) lists different possible methods of administration which the Commission may choose to use singly or in combination, but it also makes it clear that an 'appropriate' method must be used. On each afternoon of the opening days determined by the Commission, it is decided how much of the total quota is to be drawn down, and the applications which have been received but not yet met are to be dealt with in one of the appropriate administrative methods. The methods, which are clearly set out in Article 144(2) are: the 'first come, first served' method, based on the chronological order of the lodging of applications; the 'simultaneous examination method', with allocation in proportion to the quantities requested when the applications were lodged; and 'traditional/newcomers method', which takes traditional trade patterns into account.

---

188. Articles 146 and 147 lay down specific rules for individual products and individual Member States.
189. On the fundamental principle of equal treatment, see Ch. 3, §3.02[B].

[4] Safeguard Measures Against Imports of Agricultural Products

In Council Regulation (EC) No 625/2009 on common rules for imports from certain third countries and Commission Regulation (EC) No 630/2009 amending the representative prices and additional import duties for certain products in the sugar sector, the Union legislator has laid down special rules for safeguard measures against imports of agricultural products to the European Union. Article 159 of the Single CMO Regulation emphasizes that, subject to certain formal requirements, these rules also apply to a certain extent to agricultural products.[190] Regulation (EC) No 625/2009 concerns imports of goods from a number of specified countries, primarily Russia and several former Soviet republics, as well as North Korea and Vietnam.[191] Regulation (EC) No 260/2009 is concerned with imports from all other third countries for which there not specific rules, i.e., primarily other countries that are Members of the WTO.[192] In Article 1(2) of both regulations, the principle is stated that imports of agricultural products from both groups of countries are free and not subject to quantitative import restrictions. However, this starting point can be derogated from by safeguard measures where free circulation of imports causes, or threatens to cause, serious harm to Union producers.[193] In these cases, the Commission can change the import conditions for the good in question so that it can only be admitted to free circulation upon presentation of an import license issued in accordance with guidelines and within a framework established by the Commission.[194]

A number of different factors can be included in the assessment of the serious harm or threat of serious harm which may be caused to producers in the Union. Such an assessment may take account of the extent of the imports, in particular if they have increased significantly either absolutely or relative to consumption in the Union. The price of the imported goods can also affect the assessment, particularly if there is significant price undercutting compared with corresponding goods produced in the Union. Finally, account can be taken of the consequent impact on Union producers of similar or directly competing products.[195]

---

190. Article 158(1) of the Single CMO Regulation refers to Council Regulations (EC) No 519/94 and (EC) No 3285/94. Regulation (EC) No 519/94 has been repealed and replaced by Regulation (EC) No 625/2009. References to the repealed legislation, including references in the Single CMO Regulation, are to be construed as references to the current regulations; see Art. 21 of Regulation (EC) No 625/2009 and Art. 26 of Regulation (EC) No 260/2009.
191. See Annex I of the Regulation for a list of the third countries covered.
192. The WTO was established at the end of the GATT Uruguay Round. Annex IA to the Agreement Establishing the WTO contains, among other things, the General Agreement on Tariffs and Trade 1994 (GATT 1994) and the Agreement on Safeguards. The Uruguay Round also produced an Agreement on Agriculture. On the international framework for the Union agricultural policy, see Ch. 9.
193. In addition to safeguard measures, the regulations also deal with surveillance, quantitative restrictions and temporary safeguard measures.
194. See Art. 15 of Regulation (EC) No 625/2009 and Art. 16 of Regulation (EC) No 260/2009.
195. See Art. 8 of Regulation (EC) No 625/2009 and Art. 10 of Regulation (EC) No 260/2009.

[5]   Inward Processing Arrangements

Union law regulates the area of customs under Regulation (EC) No 450/2008 laying down the Community Customs Code (the 'Customs Code').[196] Among other things, the Customs Code deals with 'inward processing' which is a special customs procedure under which products can be imported into the Union with a view to their being processed, without the payment of customs or duties, on condition that the processed goods are then exported to third countries and are not put on the market in the Union. The Customs Code lays down detailed rules for the procedure for inward processing in Articles 114–129 of the Customs Code, where it is stated that it is the individual importer who initiates the procedure for one or more consignments of goods which they wish to import. However, Article 160 of the Single CMO Regulation gives the Commission authority to suspend inward processing arrangements for a number of the most important agricultural products.

[B]   Exports to Third Countries

[1]   Export Licenses

As with imports of agricultural products, the Single CMO Regulation contains rules to ensure Union control over the extent of trade with third countries where this trade takes the form of the export of agricultural products. Article 161 of the Single CMO Regulation empowers the Commission to decide to allow exports of a number products subject to presentation of an export license.[197] This relates to products in the sectors of cereals, rice, sugar, olive oil and table olives, fruit and vegetables, processed fruit and vegetables, wine, beef and veal, milk and milk products, pig meat, sheep meat and goat meat, eggs, poultry meat, and agricultural ethyl alcohol.[198] Pursuant to Article 161(2), Articles 131–133, on issuing import licenses and their validity and the provisions of security, apply *mutatis mutandis*.[199] Article 161(3) authorizes the Commission to adopt detailed rules for the application of rules on export licenses, including the terms of validity of the licenses and the rate of security.[200]

---

196. The Community Customs Code was introduced by Council Regulation (EEC) No 2913/92 establishing the Community Customs Code. Since its introduction, the Customs Code has been amended many times, and the latest codified version is in Regulation (EC) No 450/2008 which is designated the Modernized Customs Code.
197. Article 161(1) uses the wording: 'Without prejudice to cases where export licenses are required in accordance with this Regulation, the Commission may make', but in its present form the Single CMO Regulation does not contain any requirement regarding export licenses.
198. In respect of olive oil, only oil as referred to in Annex I, Part VII, point (a).
199. On Arts 131 to 133, see §4.06[A][1].
200. See Commission Regulation (EC) No 376/2008 laying down common detailed rules for the application of the system of import and export licenses and advance fixing certificates for agricultural products, and Commission Implementing Regulation (EU) No 282/2012 laying down common detailed rules for the application of the system of securities for agricultural products. See also the general comments on import licenses in §4.06[A][1].

[2] *Export Refunds: Financial Support for Exports*

In connection with imports of agricultural products, the Union uses customs duties and additional customs duties as a financial instrument to regulate the flow of goods into the internal market. In the same way, export refunds, i.e., financial support for the export of agricultural products, are used as financial instruments for regulating the flow of goods from the Union to third countries. Article 162(1) states the basic rule that: to the extent necessary to enable exports on the basis of world market quotations or prices, the difference between those quotations or prices and prices in the Union may be covered by export refunds. The rule expressly states that the right to equalize price differences etc. only applies within the limits resulting from agreements entered into pursuant to Article 218 TFEU.[201]

Not all exports of agricultural products are eligible to receive export refunds. Article 161(1)(a) and (b) distinguishes between products that are exported without further processing and products which have been subject to some processing. With regard to products without further processing, subparagraph (a) provides that export refunds can be granted for cereals, rice, certain sugar products, beef and veal, milk and milk products, pig meat, eggs, and poultry meat.[202] For the list of processed products for which export refunds may be granted, subparagraph (b) refers to Annexes XX and XXI. This involves a long list of products from the cereals, rice, sugar, milk and egg sectors.[203]

According to Article 164(2), the Commission fixes the amounts of export refunds, and Article 164(1) requires export refunds to be the same for the whole Union. In other words, exporters must be entitled to refunds of the same amounts, regardless of the Member State in which they are located. However, the Commission may differentiate between different rates for different destinations. The possibility for differentiating according to destination exists 'especially' where the world market situation, the specific requirements of certain markets, or obligations resulting from agreements concluded in accordance with Article 218 TFEU make this necessary.[204]

When the Commission sets the rate for refunds for a specific product, Article 164(3) requires it to take into account one or more of a number of named aspects. The eight general aspects (and the ninth which applies specifically to pig meat, eggs and poultry meat) are broadly formulated, and since the Commission is merely required to take these aspects 'into account', they do not set very strict limits to the Commission's discretion on how big the refunds should be. However, for products that are exported as processed products, as referred to in Annexes XX and XXI, Article 162(2) provides that the export refunds may not be higher than those that apply to equivalent products that are unprocessed.

---

201. Article 162(1) refers to Art. 300 of the EC Treaty, which corresponds to Art. 218 TFEU.
202. In respect of sugar products, this refer to products listed in points (b), (c), (d) and (g) of Part III of Annex I.
203. There is a special restriction on the processed products in the milk and milk products sector for which refunds can be made; see Art. 162(1). And there is a special rule for certain categories of spirit drinks obtained from cereals; see Art. 162(3).
204. Article 164(1) refers to Art. 300 of the EC Treaty, which corresponds to Art. 218 TFEU.

Article 167 lays down the conditions that an exporter must fulfil before an export refund can be paid. Under paragraph (6), there is a general condition that a refund can only be paid upon submission of proof that the products have been exported from the Union, and in the case of a differentiated refund, that the products have reached the specified destination.[205] However, exceptions to the requirements for documentation may be allowed by the Commission if conditions are laid down which offer equivalent guarantees.

In addition to the general conditions in Article 167(6), there are formal distinctions depending on whether the products in question are processed or unprocessed. For unprocessed products Article 167(1) provides that refunds are only granted on application and on presentation of an export license.[206] The requirement for an export license does not follow from Article 167, but paragraph (4) gives the Commission authority to decide on this.[207]

As stated, under Article 164(2) the amounts of export refunds are fixed by the Commission, either on an ongoing basis or by tendering. In this connection, Article 167(2) provides that the refund applicable to unprocessed products is that applicable on the day of application for the license or that resulting from the tender procedure concerned. A special rule applies in the case of a differentiated refund.

Article 168 makes export refunds for live cattle conditional on compliance with Union legislation on animal welfare and, in particular, the protection of animals during transport.

Under international agreements with third countries and with international organizations, the Union and the Member States have accepted limits to grants of export support in the agricultural sector. Article 169 thus provides that volume commitments resulting from international agreements must be complied with. The basis for such compliance is export licenses issued for the determined reference periods.

By its power laid down in Article 170, the Commission has adopted implementing provisions for the rules on export refunds.[208]

Articles 171, 172 and 173 contain special rules on the management of tariff quotas opened by third countries for milk and milk products, the issue of certificates for products benefiting from special import treatment in a third country, and the setting of

---

205. On refund paid to the exporter after submission of documents forged by its contracting partner, see Case C-143/07 *AOB Reuter & Co.* [2008] I-3171. On absence of stock records and proof of that the goods has been exported, see Case C-402/10 *Société Groupe Limagrain Holding* [2011] ECR.
206. For eggs for hatching and day-old chicks, the Commission may decide that export licenses may be granted *ex post*; see Art. 167(3).
207. In this case, the decision is taken in accordance with the procedure in Art. 16(2) of Council Regulation (EC) No 1216/2009, which refers in turn to Arts 5 and 10 of Regulation (EU) No 182/2011 (the examination procedure). On the committee (comitology) procedure, see above in Ch. 3, §3.05[C]. Article 167(4) of the Single CMO Regulation refers to Council Regulation (EC) No 3448/93, which was repealed and replaced by Regulation (EC) No 1216/2009; see Art. 20 of the latter. Regulation (EC) No 1216/2009 refers to Council Decision 1999/468/EC which was repealed and replaced by Regulation (EU) No 182/2011; see Arts 12 and 13 of the latter.
208. See Commission Regulation (EC) No 612/2009 on laying down common detailed rules for the application of the system of export refunds on agricultural products.

minimum prices for exports of certain products in the live plants sector to third countries.

[3]   Outward Processing Arrangements

The procedure for inward processing and the background to it are described above in section §4.06[A][5]. The counterpart to this is the procedure for outward processing. Under outward processing, agricultural products produced in the Union are temporarily exported with a view to processing outside the Union and then, either wholly or partly exempt from customs and duties, re-imported to the Union. The Customs Code in Regulation (EC) No 450/2008 lays down a number of rules for outward processing.[209] According to these, permission for outward processing can be granted to the economic operator who wants the processing carried out. However, Article 174 of the Single CMO Regulation gives the Commission authority to suspend the use of outward processing arrangements for the products of the cereals, rice, fruit and vegetables, processed fruit and vegetables, beef and veal, pig meat, sheep meat and goat meat and poultry meat sectors. Such suspension may occur if the Union market is disturbed or is liable to be disturbed by outward processing arrangements.

[C]   Force Majeure and Forfeiture of Security Lodged in Connection with Licenses Etc.

Many of the rules of Union agricultural law impose strict formal requirements on farmers and the agricultural sector's processing and distribution undertakings. Under Union law, the failure to comply with such requirements can trigger the imposition of heavy sanctions by the authorities. Where it is established that an agricultural undertaking has not fulfilled the conditions of Union law, and where a sanction is attached to the failure to fulfil the requirement, individual undertakings can try to avoid the sanctions by relying on the fundamental principle of proportionality in Union law. In a number of cases and with varying degrees of success, undertakings have argued that the sanctions laid down go further than is necessary and appropriate for the lawful purposes for implementing the sanctions.[210]

If the legislative act with its sanctions is regarded as lawful, then farmers and other undertakings in the agricultural sector may be able to claim force majeure. Claiming force majeure generally involves the agricultural undertaking referring to wholly unusual circumstances that lie outside the undertaking's ability to influence and which have occurred despite the undertaking having taken all the safeguard measures that can reasonably be expected of a prudent and conscientious economic

---

209. See Regulation (EC) No 450/2008 on the Customs Code.
210. See Case 11-70 *Internationale Handelsgesellschaft* [1970] ECR 1125; Case 122/78 *Buitoni* [1979] ECR 677; Case 137/85 *Maizena* [1987] ECR 4587 and Case C-339/92 *ADM* [1993] ECR I-6473. On the application of the proportionality principle in Union agricultural law, see Ch. 3, §3.02[C].

operator.[211] The effect of force majeure is that the authorities cannot impose sanctions on the agricultural undertaking, even if the undertaking has not fulfilled its obligations under the legislation.

Force majeure has particularly been claimed in connection with Union law arrangements for the provision of security in connection with grants of import and export licenses. As stated in sections §4.06[A][1] and §4.06[B][1] above, Union law generally requires a license for imports or exports of agricultural products to or from countries outside the Union. Previously these arrangements were regulated by numerous market organizations for individual agricultural products, but they are now regulated by the Single CMO Regulation. Articles 130–134 and 161 contain the rules on the provision of security in connection with the grant of import and export licenses.

[1]   The Term Force Majeure in Union Agricultural Law

The interpretation of the term 'force majeure' in Union agricultural legislative acts has given rise to a number of cases before the Court of Justice.[212] In the *Kampffmeyer* case, the Court of Justice stated that the concept of force majeure differs in content in different areas of the law and in its various spheres of application, so that the precise meaning of the concept has to be decided by reference to the legal context in which it is used.[213] In the *Internationale Handelsgesellschaft* case, the Court of Justice had already stated that: 'the concept of force majeure adopted by the agricultural regulations takes into account the particular nature of the relationships in public law between traders and the national administration, as well as the objectives of those regulations'.[214] This view has been upheld in more recent cases meaning that one can confidently assume that the force majeure concept is the same in cases concerning import and export licenses for agricultural products.[215]

This basic view presumably also means that the definition of the concept of force majeure is the same in other areas of agricultural law, where the provision of security is required in order to benefit from a license or some other advantage.[216] In the

---

211. For further on the content of the concept of force majeure, see §4.06[C][1] below.
212. Previous Union legislation in the area of agriculture exhaustively listed the events that could be considered force majeure; see e.g., Regulation 87/1962 'relatif à l'établissement des modalités d'application concernant les certificats d'importation et d'exportation pour les céréales et les produits céréaliers', JO 66 du 28.7.1962, p. 1895.
213. Case 158/73 *Kampffmeyer* [1974] ECR 101, para. 8.
214. Case 11/70 *Internationale Handelsgesellschaft* [1970] ECR 1125, para. 23. The term 'agricultural regulations' must be read as a general reference to agricultural legislation and not merely to the rules that applied in the particular case; in the French and German language versions the terms 'les reglements agricoles' and 'im Bereich der Agrarverordnungen' are used respectively. See also Case 4/68 *Schwarzwaldmilch* [1968] ECR English special edition 377.
215. For a restatement of the basic position referred to above, see Case 25/70 *Köster* [1970] ECR 1161, para. 38; and Case 158/73 *Kampffmeyer* [1974] ECR 101, para. 17, where the Court of Justice referred to 'the special nature of the relationships at public law existing between the importers and the national administration'.
216. Gilsdorf, 'La force majeure dans le droit de la CEE a la lumiere de la jurisprudence de la Cour de justice', CDE 137 (141) (1982), argued that force majeure can mean different things in different contexts and is not a general principle: 'Il est évident que le concept de la "force

Denkavit France case, the Court of Justice stated that the concept of force majeure must be assessed on the basis of the provisions in each of the regulations in which the term is used.[217] But this appears too categorical, and from paragraphs 28 and 29 of the judgment one can get the impression that the statement covers a mixture of the legal definition of the term and an assessment of the actual facts in the case.

[2]   The Content of the Force Majeure Concept

In the *Internationale Handelsgesellschaft* case, which concerned the loss of security paid in connection with an export license under the then market organization for cereals, the Court of Justice stated in general on the interpretation of the concept of force majeure that:

> the concept of *force majeure* is not limited to absolute impossibility but must be understood in the sense of unusual circumstances, outside the control of the importer or exporter, the consequences of which, in spite of the exercise of all due care, could not have been avoided except at the cost of excessive sacrifice.[218]

In two subsequent cases on the provision of security in connection with licenses for the import of frozen beef and barley, the Court of Justice mitigated the requirements for licensees as, in addition to the requirement for unusual circumstances, outside the control of the licensee, it now stated that the concept of force majeure is not limited to cases of absolute impossibility but includes abnormal circumstances outside the control of the importer, which have arisen in spite of the fact that the licensee 'has taken all the precautions which could reasonably be expected of a prudent and diligent trader'.[219]

---

majeure" ne constitue pas un principe général de droit mais traduit une situation de fait exceptionnelle qui exclut les conséquences juridiques normale d'un dispositif réglementaire'; while Flynn, 'Force Majeure Pleas in Proceedings Before the European Court', European L. Rev. 102 (114) (1981), argued that force majeure is a general principle of Union law, in the form of a subset of the proportionality principle. See also Loyant, 'La force majeure et l'organisation commune des marchés agricole', RTDE 255 (1980); and Schulze, *Ratgeber Agrarreform* 86 and 94-96 (Landwirtschaftsverlag, Ed. 2, 1998). On force majeure in connection with trade with Comecon countries (the Soviet block prior to the fall of the Iron Curtain), 'state trading countries', see Sośniak, 'Staatsakt und höhere Gewalt im internationalen Handelsverkehr der RGW-Länder', RIW 105 (1984).

217. Case 266/84 *Denkavit France* [1986] 149, para. 27.
218. Case 11/70 *Internationale Handelsgesellschaft* [1970] ECR 1125, para. 23.
219. Case 186/73 *Norddeutsches Vieh- und Fleischkontor* [1974] ECR 533, para. 7; and Case 3/74 *Pfützenreuter* [1974] ECR 589, para. 8. Usher, *EC Agricultural Law* 88 (Oxford University Press, 2001), states that in these two cases the Court of Justice 'slightly relaxed' the definition of force majeure. However, in contrast, in Commission notice C(88) 1696 concerning force majeure in European agricultural law (OJ C 259, 6 Oct. 1988, 10), the Commission stated in para. 1 (together with footnote 6) that the definition given in the *Internationale Handelsgesellschaft* case had been repeated by the Court in a large number of judgments, not only concerning agriculture but also other spheres, and that the varying forms of words used by the Court in its decisions must be regarded as resulting from the specific characteristics of the cases in question.

117

[3]   The Force Majeure Decision in the Specific Case

When deciding on force majeure in a specific case, it is important to be aware of the division of responsibilities between the Court of Justice and the national courts. It is only the interpretation of the concept of force majeure under Union law that falls under the jurisdiction of the Court of Justice, while its application to the facts of the case is a matter for the national courts. In the *Kampffmeyer* case, the Court of Justice ruled that the question of whether the commercial undertaking had acted as a prudent and diligent trader was not a matter of interpretation, but a matter of the application of the law which must be decided by the national court, given all the facts of the situation in which the trader found itself.[220] In the *Parras Medina* case, the Court of Justice divided the tasks between itself and the national court so that it is a matter for the national court to examine whether the facts adduced by the commercial operator are correct and whether the operator has taken the necessary care to comply with the deadline laid down.[221]

Regardless of that the division of tasks laid down by the Court of Justice in the *Kampffmeyer* case is right in principle; the Court has had to acknowledge that in many cases in practice the application of the law is difficult to distinguish from the evaluation of the evidence.

In the *Kampffmeyer* case, a German authority which administered import licenses for cereals issued a license to Kampffmeyer. The license expired on 31 October 1972. In accordance with the terms of the license, Kampffmeyer arranged with a number of operators to import several consignments of cereals, but after 25 October the license was apparently lost in the post between two of these operators. The operator sending the license was located in the Netherlands and the operator receiving it was located in Germany. As result, Kampffmeyer did not succeed in importing the last consignment in time, and thus only partially used its license. The German authority then declared that part of the security provided by Kampffmeyer was to be surrendered. The Court of Justice left the evaluation of the facts to the national court, the Verwaltungsgericht in Frankfurt am Main, which found that Kampffmeyer could claim force majeure. It was not Kampffmeyer's fault that the license that disappeared had been sent by ordinary post, and it could not be a requirement that the license should be sent via a messenger. The German court reasoned that it was in accordance with normal practice in both the Netherlands and Germany that such licenses should be sent via the ordinary post, and stated that sending a letter by registered post did not give a guarantee that it would not be lost.[222]

In a number of cases, the Court of Justice has shown itself willing to assess whether a given situation falls within the force majeure concept that is used in the agricultural regulations. The *Corman* case concerned Regulation (EEC) No 232/75, which gave undertakings the right to bid to buy butter at reduced prices for processing

---

220. Case 158/73 *Kampffmeyer* [1974] ECR 101, para. 27.
221. Case C-208/01 *Parras Medina* [2002] ECR I-8955, para. 22.
222. Unpublished judgment of 20 Mar. 1974 of the Verwaltungsgericht Frankfurt/Main, (II/V – – E 48/73), pp. 6-8.

in the manufacture of pastry products and ice-cream within a limited period. If the processing did not take place within the set period, the undertaking would lose the security it had provided in whole or in part. The Belgian company Corman bought a consignment of butter for processing and lawfully sold it to an English company. The British customs authorities had issued a customs document as evidence that the processing had taken place. However, subsequent monitoring showed that Corman's buyer had slipped up in connection with selling the butter on, and had not ensured that the processing of the butter took place within the period allowed. In arguing that it should not lose the security provided, Corman 'relied on a customs officer who purported to have authority to grant an extension' to Corman's buyer, and the fact that it subsequently appeared that the customs officer did not have the authority to grant an extension to the buyer constituted force majeure with respect to Corman. The Court of Justice rejected this argument, stating that a successful tenderer such as Corman had several means available to it by which it could have been indemnified if a third party did not act so as to ensure the correct and timely processing of the butter. For example, the Court referred to the possibility of Corman requiring the provision of security or the inclusion of a compensation clause in the contract under which Corman sold the butter to a third party.[223] In these examples, the Court of Justice has not considered it important that a party to a contract does not always have the negotiating strength enabling them to include such clauses in their contracts.

In the *Jaka* case, Jaka was a company manufacturing and exporting tinned ham, and it lost its ability to produce and thus export its goods because of strikes in other undertakings responsible for transport and cleaning. This raised the question of whether a strike in an undertaking other than the licensee could be regarded as force majeure. The Court of Justice ruled that such a strike must normally be regarded as a circumstance over which the licensee has no influence, as it did not have any possibility of influencing the circumstances that have led to the strike. However, the Court did not find that a strike held following lawful notice which could affect the licensee's activities could be regarded as an abnormal or unforeseeable circumstance.[224]

---

223. Case 125/83 *Corman* [1985] ECR 3039, paras 29–31. In Case 109/86 *Theodorakis* [1987] ECR 4319, the Court of Justice ruled that the licensee could not claim force majeure on the ground that its counterparty was in breach of an agreement to export the goods in question.
224. Case C-338/89 *Jaka* [1991] ECR I-2315, paras 17–25. The voluntary transfer of a farm cannot be regarded as force majeure; see Case C-376/97 *Wettwer* [1999] ECR I-3449, para. 30, on the loss of a special premium paid to producers of beef. On the failure to export because of changed quality standards in a third country that is a state trading entity (the Soviet Union), see Case C-124/92 *An Board Bainne* et al. [1993] ECR I-5061. In Case C-109/95 *Astir* [1997] ECR I-1385, it was established that a shipwreck in which a cargo of wheat flour was lost was force majeure. Likewise, in Case C-299/94 *Anglo Irish Beef* [1996] ECR I-1925, it was established that a trade embargo adopted by the EU's Council of Ministers in connection with the First Iraq War (Iraq's invasion of Kuwait), which affected exports of boned beef, was force majeure. In Case C-208/01 *Parras Medina* [2002] ECR I-8955, the Court of Justice ruled, on the setting aside of the obligation to report on a farm's stock of wine, that the death of the sole manager of a family business run in the form of a mutual company with the manager being closely related to the other mutual owners, was force majeure. Damage to a consignment of beef in transit was not considered to be force majeure in Case C-218/09 *SGS Belgium NV* [2010] ECR I-2373.

[4]   *Summary of Force Majeure in Case Law and Contractual Possibilities*

While it is possible to trace some development of the Court of Justice's definition of force majeure in agricultural cases, the definition must still be characterized as being strict. There have been few cases in which agricultural undertakings have been successful in claiming force majeure against the authorities. On this, the Court's case law on force majeure is in line with its practice on the application of the proportionality principle to sanctions for breaches of the rules of agricultural law.[225]

If a case is pending before the national courts and the question is raised as to whether force majeure exists, the parties can ask for the issue to be referred to the Court of Justice for a preliminary ruling pursuant to Article 267 TFEU. Such request should take into account that the definition of force majeure, including the abstract assessment of the facts, is well established in the practice of the Court, and that the national courts *may* adopt a milder assessment of the concrete facts.

In its decisions, the Court of Justice has pointed out that individual undertakings can, to a certain extent, protect themselves against abnormal circumstance that do not fall within the ambit of force majeure by contractual provisions. To the extent that one undertaking is dependent on another undertaking for the fulfilment of its obligations under Union law, where commercially feasible the first undertaking should contractually clarify the liability for any loss (and the provision of security) which it might incur through sanctions imposed under the law.

§4.07   THE RULES ON COMPETITION AND STATE AID

According to Article 42 TFEU, the Treaty's chapter relating to rules on competition only apply to the production of and trade in agricultural products to the extent determined by the European Parliament and the Council. The Union legislator has exercised this power in the adoption of both Council Regulation (EC) No 1184/2006 of 24 July 2006 applying certain rules of competition to the production of, and trade in, agricultural products and Articles 175–182 of the Single CMO Regulation. These rules are examined in Chapter 8.

§4.08   SPECIAL PROVISIONS FOR INDIVIDUAL SECTORS AND
GENERAL PROVISIONS

Part V, Articles 183–189, of the Single CMO Regulation contains special provisions for individual sectors. This is a mixed bag of rules that the legislator could not find an appropriate place for elsewhere in the Regulation.

Article 183 authorizes Member States to impose a promotional levy on its milk producers in respect of marketed quantities of milk or milk equivalent in order to finance measures promoting consumption in the Union, expanding the markets for milk and milk products and improving quality.[226] This authority can only be exercise

---

225. On the proportionality principle, see Ch. 3, §3.02[C].
226. On support measures for milk and milk products, see above in §4.04[D][3].

within the framework set out in Article 180 of the Regulation on the application of the TFEU State aid rules within the scope of the Single CMO Regulation.[227]

Under Article 184, the Commission is bound to present a report to the European Parliament and the Council on a range of agricultural sectors and matters associated with them within certain deadlines. These concern the dried fodder, apiculture, the German alcohol monopoly, the fruit and vegetable sector, the school fruit scheme, the milk quota arrangements, circumstances related to cheese producers with protected designations of origin, sales promotion measures to third countries and experiences of reform of the wine sector.

There are special rules for the hops sector, the wine sector and the bioethanol sector in Articles 185, 185a to 185d and 189. The rules for the hops sector concern an obligation to register contracts within the sector, while the rules for the bioethanol sector require Member States to inform the Commission about certain matters relating to the sector.

Articles 186–188 give the Commission authority to take the necessary measures to deal with disturbances in a number of important sectors.

Part VI of the Single CMO Regulation contains general provisions. Article 191 allows the Commission to adopt measures that are both necessary and justifiable in an emergency in order to resolve specific practical problems. Such measures may derogate from provisions of the Regulation, but only to the extent and for such a period as is strictly necessary. In addition, Articles 192 and 194 give the Commission authority to lay down more detailed provisions on the exchange of information between the Member States and the Commission, and controls, administrative measures and administrative penalties for breaches of the provisions of the Regulation.

There is a special rule in Article 193. Unless otherwise specifically provided, no advantage provided for under the Single CMO Regulation may be granted to a natural or legal person if it is established that the conditions for obtaining such an advantage has been created artificially, contrary to the objectives of the Regulation.[228]

---

227. See §4.04[D][6] above where there is a reference to Ch. 8.
228. See Ch. 3, §3.02[D], on the fundamental principle of the protection of legitimate expectations.

CHAPTER 5
Structural and Rural Development Aid

§5.01 THE CONTENT AND DEVELOPMENT OF STRUCTURAL AND RURAL DEVELOPMENT POLICY

The structural policy of the agricultural sector has traditionally been defined as covering all measures which, in the medium or long term, will change the structures for the production, processing, sale and consumption of agricultural products.[1] However, this definition is no longer suitable to cover the latest developments in rural development policy. While in the past structural policy has been linked in some form or other to agricultural production, rural development policy has broader aims. Rural development policy also covers measures concerning people who live in rural areas but who are not engaged in agriculture. It can be said that for traditional structural policy what mattered was what people or undertakings did, whereas for rural development policy what matters is where people live or undertakings are based, whether or not they are involved in agricultural production.

Traditional structural policy measures can either take the form of rules or support schemes. In relation to the regulation of structures, this can concern the legal framework within which production can take place, including forms of ownership and rules on the leasing of farms, rules on production itself, including environmental regulations, rules on quality and on the manufacture of feeding stuffs, veterinary rules, rules on the health of plants and animals, rules on the marketing of agricultural products, and rules on market conduct, such as competition rules and rules on producer groups. Traditional structural policy measures in the form of support schemes can take the form of training schemes for people engaged in agriculture, the modernizing of farms or the development of infrastructure.

The structures of supply and demand in the area of agriculture depend on a range of factors. First, there are natural factors, such as climate, soil quality and characteristics, and the availability of water in general. Second, there are demographic factors in

---

1. Barents, *The Agricultural Law of the EC* 173 (Kluwer, 1994).

the form of growing or diminishing populations. Third, there is the level of the production apparatus, such as the number and sizes of farms, the productivity and mechanization of farms, the professional skills of farmers, and the infrastructure in general. Finally, the social and legal conditions, such as competition, the leasing of land and property, tax, the labour force and general environmental requirements can all be relevant factors. Change to these factors is generally something that takes place in the medium term (one to three years) or the longer term (three to five years).[2] Within the EU, agriculture is characterized by the existence of very wide differences between these many factors, and together they form the structure of the sector. These differences were not diminished by the entry of the Central and Eastern European Member States, and in themselves they underline the need for legislation at Union level.[3]

Article 39 of the TFEU sets out the purposes and considerations of the common agricultural policy, and it is usually emphasized that the structural policy is provided for in both paragraphs (1) and (2). Article 39(1)(b), expresses the purpose of ensuring a fair standard of living for the agricultural community. This is supported by the consideration in Article 39(2) that account shall be taken of the particular nature of agricultural activity, which results from the social structure of agriculture and from structural and natural disparities between the various agricultural regions.[4]

In Article 42, second paragraph, TFEU it is stated that, under certain conditions, the Council may authorize the granting of aid for the protection of enterprises handicapped by structural conditions.[5]

For a long time, structural policy at EU level has had a lower priority than price and market policies, as expressed in the market organizations and now in the Single CMO (common market organization) Regulation. In recent years, this prioritizing has been changed somewhat.[6] In Agenda 2000, structural policy in the agricultural sector was gathered together in the framework Council Regulation (EC) No 1257/1999 on support for rural development from the European Agricultural Guidance and Guarantee Fund (EAGGF), whose provisions are largely carried forward in Council Regulation (EC) No 1698/2005 on support for rural development by the European Agricultural Fund for Rural Development (EAFRD) (the 'Rural Development Regulation').

---

2.   *Ibid.*
3.   See Cardwell, 'Rural Development in the European Community: Charting a New Course?', 21 Drake J. Agric. L, (2008), on the necessity for Union legislation on rural development.
4.   This consideration is particularly emphasized in recital 2 of Council Regulation (EC) No 1698/2005 on support for rural development by the European Agricultural Fund for Rural Development (EAFRD).
5.   On State aid, see Ch. 8, §8.04.
6.   See Ch. 1, §1.01, on the development. In particular on the development of structural policy from the Stresa Conference and onwards see Barents, *The Agricultural Law of the EC* 174–178 (Kluwer, 1994); McMahon, *EU Agricultural Law* 65–71 (Oxford University Press, 2007); and Cardwell, 'Rural Development in the European Community: Charting a New Course?', Drake J. Agric. L. 21 (24-34) (2008). In 1982 Constantinides-Megret, *La politique agricole commune en question,* 167 (A. Pedone, 1982), could state that: '*L'organisation commune des marchés agricoles est principalement fondée sur régime de prix communs garantis.*'

§5.02 THE RURAL DEVELOPMENT REGULATION

The Rural Development Regulation contains five main groups of rules which are defined in its Article 1. Article 1(1) states that the Regulation lays down the general rules governing Union support for rural development. This is financed through the EAFRD. The EAFRD was set up by Council Regulation (EC) No 1290/2005 on the financing of the common agricultural policy. According to Article 1(2) of the Rural Development Regulation, the second main group of rules defines the Objectives to which rural development policy is to contribute. According to Article 1(3), the third group of rules defines 'the strategic context for rural development policy', including the method for fixing the Union's strategic guidelines for rural development policy and the national strategy plans. According to Article 1(4), the fourth group defines the priorities and measures for rural development. And finally, according to Article 1(5), the fifth group lays down rules on the sharing of administrative responsibilities between the Member States and the Commission.

On the face of it, the second and fourth groups of rules, referred to in Article 1 (2) and (4) are the same. It can be difficult to distinguish between objectives and priorities.[7] However, the use of the word 'Objectives' in Article 1(2) emphasizes that this is a reference to the Objectives specified in Article 4.[8] The words 'priorities and measures' in Article 1(4) is a reference to the four axes, for which specific rules are set out in Articles 20–65. This explanation of Article 1(4) means that the provisions should read as saying that: 'This Regulation ... defines *the Axes* (i.e., priorities and measures) for rural development'.[9]

The policy and its rules for rural development do not stand alone. Article 3 makes it clear that the financing of developments through the EAFRD supplements the market and income support policies of the common agricultural policy. Today this is a reference to the Single CMO Regulation.[10]

[A] The Objectives of the Rural Development Policy and Its Four Axes

The overall objective of the rural development policy is the development of rural areas. In the preamble to the Rural Development Regulation, it is stated that this objective cannot be achieved sufficiently by the Member States given the links between rural

---

7. In the French language version, the terms used are '*les objectifs*' and '*les priorités*', while in the German language version the terms used at '*die Ziele*' and '*die Schwerpunkte*' *(Prioritäten)*.
8. On the objectives of rural development policy, see below in §5.02[A].
9. In the German language version, Art. 1(4) uses the term '*schwerpunkte*', and the same term is used in the heading for Title IV, Ch. I. In the French language version, Art. 1(4) uses the wording '*les priorités et les mesures*', and in the heading for Title IV, Ch. I, it uses the word 'Axes'. On the strategic guidelines, strategy plans and programmes, see §5.02[B] below, and on the four axes, see §5.02[A].
10. For a criticism of the relationship between the goals under the Single CMO Regulation and the goals under the rural development policy, see Bodiguel, 'The New European Rural Development Regulation: Implementation in France', Drake J. Agric. L. 7 (2008). For a criticism of the rural development policy with an emphasis on the limited financial prioritizing, see Cardwell, 'Rural Development in the European Community: Charting a New Course?', Drake J. Agric. L, 21 (2008).

development and the other instruments of the common agricultural policy, and due to the extent of the disparities between the various rural areas of the Member States. The objective can therefore be better achieved at Union level through the multiannual guarantee of Union finance.[11]

Article 4(1) specifies the three overall objectives which, according to Article 4(2), are to be implemented by means of the four axes defined in Title IV, i.e., Articles 20–65.[12]

Under Article 4(1)(a), the first objective support for rural development is to contribute to improving the competitiveness of agriculture and forestry by supporting restructuring, development and innovation. This objective is to be achieved by means of axis 1, which is described in detail in Articles 20–35a. This axis includes a number of measures that lie at the heart of traditional structural policy and which are carried forward from previous legislation.[13] The measures are all linked to production in the agriculture and forestry sectors and can concern aid for the establishment of young farmers, the modernization of farms, improvement of the value of forests or aid for producer groups with respect to information and marketing activities for products that are covered by measures on food quality.[14]

The second objective, Article 4(1)(b), is the improvement of the environment and the countryside by supporting land management. This objective is to be achieved by means of axis 2, set out in Articles 36–51 under the heading of 'Improving the environment and the countryside'. This axis includes measures on the sustainable use of agricultural land and forestry land. Such measures can take the form of payments for environmentally-friendly agriculture and forestry, payments for animal welfare, and aid for non-productive investments. These latter can be for the enhancement of the public amenity value of agricultural and wooded areas, as referred to in Article 41(b) and Article 49(b).[15]

The third objective, referred to in Article 4(1)(c), is improving the quality of life in rural areas and encouraging diversification of economic activity. This objective is to be achieved by axis 3, set out in Articles 52–60. These include measures to develop local infrastructures and human resources in rural districts, among other things by developments in sectors other than agriculture.[16] It is particularly by measures under this axis that the framework for traditional structural policy has been expanded.[17]

In addition to the achievement of the three objectives, each under its own axis, Articles 61–65 of the Rural Development Regulation contain axis 4, which incorporate the 'Leader approach'. The aim of this method is to ensure as high a level as possible of local influence on the measures put in place in the rural districts in question. This is

---

11. See recital 5 of the Rural Development Regulation.
12. See Cardwell, 'Rural Development in the European Community: Charting a New Course?', Drake J. Agric. L. 21 (2008).
13. See e.g., Council Regulation (EC) No 1257/1999 on support for rural development from the European Agricultural Guidance and Guarantee Fund (EAGGF).
14. See §5.02[A][1] on the rules in axis 1.
15. See §5.02[A][2] on the rules in axis 2.
16. See recital 46 of the Rural Development Regulation.
17. See §5.02[A][3] on the rules in axis 3.

to be done by establishing local action groups to draw up strategies for the development of the local rural district or area. According to Article 63(a), these local development strategies should aim to achieve the objectives of one or more of the three other axes. Council Decision 2006/144/EC on Community strategic guidelines for rural development (programming period 2007 to 2013), refers to axis 4, based on the Leader approach, introducing possibilities for 'innovative governance through locally based, bottom-up approaches to rural development'.[18] Among other things, axis 4 with its Leader approach is based on the Commission White Paper on European governance, which states that: 'At EU level, the Commission should ensure that regional and local knowledge and conditions are taken into account when developing policy proposals.'[19]

[1]   *Axis 1: Improving the Competitiveness of the Agricultural and Forestry Sector*

Support for the development of rural districts under axis 1, i.e., support for improving the competitiveness of the agricultural and forestry sector, is divided into three groups. According to Article 20, first there can be measures aimed at promoting knowledge and improving human potential. Second, there can be measures aimed at restructuring and developing physical potential and promoting innovation. And finally, there can be measures aimed at improving the quality of agricultural production and products.[20]

For the first group, these very broad definitions of the various measures are specified in Article 20(a)(i) to (v), and in Articles 21–25. The group includes support measures for vocational training and the diffusion of scientific knowledge, support for the setting up of young farmers, early retirement of farmers and farm workers, and the setting up and use of various forms of farm management, farm relief and farm advisory services.

The second group of measures, aimed at restructuring and developing physical potential and promoting innovation, is listed in Article 20(b)(i) to (vi) and Articles 26–30, and relate to the modernization and improvement of agricultural holdings and forests; adding value to products; cooperation for developing new products, processes and technologies in the agriculture and food sector and in the forestry sector; improving and developing the infrastructure for the development of agriculture and forestry; and restoring agricultural production potential damaged by natural disasters.

According to Article 20(c) and Articles 31–35, under the third group of measures under axis 1, support can be given for improving the quality of agricultural production and products. This relates to support to farmers to cover the costs and loss of income resulting from Union legislation on environmental protection, public health, animal

---

18. See para. 2.3 of the Decision, and see below in §5.02[A][4].
19. COM(2001) 428 final, 13.
20. Axis 1 contains a fourth group, concerning support for semi-subsistence agricultural holdings undergoing restructuring and for setting up of producer groups on the basis of a reform of a common market organization; see Art. 20(d). On axis 1 in general, see Cardwell, 'Rural Development in the European Community: Charting a New Course?', Drake J. Agric. L. 21 (40–41) (2008).

and plant health, animal welfare and occupational safety. Under this heading, support can also be given to farmers who participate in food-quality schemes.

[2]   Axis 2: Support for Improving the Environment and the Countryside

Measures under axis 2 are aimed at supporting the sustainable use of agricultural and forestry land.[21] The measures are divided into two groups, one addressed to agriculture and the other addressed to forestry.

The support for sustainable agriculture relates particularly to payments given to farmers in areas with handicaps such as mountainous areas, and payment connected to Natura 2000 and to Directive 2000/60/EC establishing a framework for Community action in the field of water policy. The Natura 2000 Directives identify a number of areas of natural interest which are deemed to be specially valuable in a European context. The aim is to ensure or re-establish favourable conservation status for natural environments and species which the areas are dedicated to preserve.[22] The agricultural measures also include agri-environment payments, animal welfare payments and support for non-productive investments. According to Article 36(a)(i) to (vi), and Articles 37–41, the latter investments include the costs for meeting environmental targets for agriculture and enhancing the public amenity value of agricultural areas.[23]

Pursuant to Article 36(a)(i) to (vii), support for the sustainable use of forestry land can relate to the conversion of agricultural land to forestry use, the first afforestation of non-agricultural land, measures associated with Natura 2000, payments for environmentally-friendly forestry practices, support for non-productive investments, and support for restoring forestry potential and introducing preventive actions.[24] This last-named aid relates to the re-establishment of forested areas that have been damaged by natural disasters and fire, and actions to prevent fires in certain forests.

Articles 50 and 51 lay down some general provisions relating to measures under axis 2. Only certain geographic areas are eligible, and certain schemes require the recipient of support to comply with conditions under Council Regulation (EC) No 1782/2003 establishing common rules for direct support schemes under the common agricultural policy and establishing certain support schemes for farmers.

---

21. Sustainability is an important part of rural development; see Cardwell, 'Rural Development in the European Community: Charting a New Course?', Drake J. Agric. L, s. 21 (22) (2008), and see 42–44 of this article on axis 2.
22. The Natura 2000 Directives are first and foremost Council Directive 92/43/EEC on the conservation of natural habitats and of wild fauna and flora; and Directive 2009/147/EC on the conservation of wild birds. See also Directive 2000/60/EC establishing a framework for Community action in the field of water policy.
23. See Art. 51 on the reduction or exclusion from support payments for failure to comply with the obligatory requirements in Council Regulation (EC) No 1782/2003 establishing common rules for direct support schemes under the common agricultural policy and establishing certain support schemes for farmers.
24. See above on Natura 2000 payments and support for non-productive investments.

[3] Axis 3: Support Improving the Quality of Life in Rural Areas and Diversification of the Rural Economy

Axis 3 gathers together arrangements that support the quality of life in rural areas and the diversification of the rural economy. This axis thus contains three of the arrangements whereby the modern rural development policy differs from the traditional structural policy for agriculture.[25] Article 52 provides that aid can be given for four groups of measures. The first group, referred to in Article 52(a)(i) to (iii) and Articles 53–55, concerns support for measures to diversify the rural economy. These can consist of diversification (presumably from agricultural activities) into non-agricultural activities.[26] Next, diversification can occur through support for the creation and development of micro-enterprises to promote entrepreneurship and develop the structure of the economy. The Regulation does not define 'micro-enterprises', but in Article 54 there is a reference to micro-enterprises as defined in Commission Recommendation 2003/361/EC concerning the definition of micro, small and medium-sized enterprises.[27] Finally, diversification can be supported by measures to promote tourism.

The second group of support measures, under Article 52(b)(i) to (iii) and Articles 56–57, concerns measures to improve the quality of life in rural areas. Under this heading, support can be given for basic services for the economy and rural population. According to Article 56, these measures can include cultural and leisure activities in a village or group of villages, and related small-scale infrastructure.[28] Support can also be given for village renewal and development. Finally, this group includes support for conservation and upgrading of the rural heritage. According to Article 57, this last-named support can include drawing-up protection and management plans relating to Natura 2000 sites and other places of high natural value.[29] There can also be environmental awareness actions and investments associated with the maintenance, restoration and upgrading of the natural heritage and with the development of high natural value sites. Other measures that can be covered by Article 57 are studies and investments associated with maintenance, restoration and upgrading of the cultural heritage such as the cultural features of villages and the rural landscape.

Article 52(c) and Article 58 contain the third group of measures. These consist solely of training and information for economic actors operating in the fields covered by axis 3. It is specified in Article 58 that these measures do not include courses of instruction or training which form part of normal education programmes or systems at secondary or higher levels.

---

25. See Cardwell, 'Rural Development in the European Community: Charting a New Course?', Drake J. Agric. L. 21 (44) (2008) on axis 3.
26. See the French language version of Art. 52(a)(i) 'vers des activités non agricoles'; and the German 'hin zu nichtlandwirtschaftlichen Tätigkeiten'.
27. OJ L 124, 20 May 2003, 36.
28. The limited size of the infrastructure does not apply to special measures pursuant to Art. 16a(1)(g) on broadband infrastructure in rural areas.
29. See §5.02[A][2] above on the Natura 2000 Directives.

The last group of support measures in axis 3 are listed in Article 52(d) and Article 59, and relate to 'a skills-acquisition and animation measure with a view to preparing and implementing a local development strategy'. According to Article 59, this kind of support can take several forms. There can be support for studies of the area concerned. There can be some doubt about what is meant by 'the area concerned'. Since Article 59 specifies the forms of support that are named in Article 52(d), and since Article 52(d) concerns measures taken with a view to preparing and implementing a local development strategy, Article 59's reference to 'the area concerned' presumably refers to support for studies for a (possible) local development strategy. Next, there can be support for measures to disseminate information about the area and the local development strategy. A third kind of measure for which aid can be given is the training of staff involved in preparing and implementing a local development strategy. A fourth kind is for promotional events and the training of leaders. Finally, under Article 52(d) and Article 59, with the exception of PR functions and further education of people with management functions, support can be given for implementing the local development strategy by public-private partnerships.[30]

[4]   Axis 4: Support Measures in Connection with the Leader Approach

As stated above in section §5.02[A], the aim of the Leader approach is to ensure as high a level as possible of local influence on the measures that are put in place in the rural districts in question.[31] The approach is defined in Article 61 as consisting of at least seven elements. First, there must be area-based local development strategies for well-identified sub-regional rural territories. Next, there must be 'local action groups' in the form of local public-private partnerships. Third, there must be a bottom-up approach, with decision-making power for local action groups on the elaboration and implementation of local development strategies. Fourth, the approach has multi-sectoral design with implementation of the strategy based on interaction between actors and projects of different sectors of the local economy. The fifth and sixth elements concern the implementation of innovative approaches and of cooperation projects. And the seventh and last element concerns networking of local partnerships.

Local action groups play a key role in the Leader approach and thus for axis 4. Article 62 lays down a number of requirements for these groups. According to Article 62(1)(b), they must consist of either a group already qualified for the Leader II or Leader+ initiatives, or according to the Leader approach, or be a new group representing partners from various locally based socio-economic sectors in the territory concerned.[32]

---

30. According to Art. 59(e), public-private partnerships may not be a partnership as defined by Art. 62(1)(b). See §5.02[A][4] below on the Art. 62(1)(b) groups.
31. See Cardwell, 'Rural Development in the European Community: Charting a New Course?', Drake J. Agric. L 21 (45) (2008) on axis 4.
32. See Notice to the Member States laying down guidelines for global grants or integrated operational programmes for which Member States are invited to submit applications for assistance in the framework of a Community initiative for rural development (Leader II) (OJ C 180, 1 Jul. 1994, 48); and Commission notice to the Member States of 14 Apr. 2000 laying down

As stated in Article 62(1)(a), the most important tasks of the local action groups is to propose an integrated local development strategy based at least on the elements contained in the definition of the Leader approach and, according to Article 62(4), to choose the projects to be financed under the strategy.

Under axis 4, pursuant to Article 63, support can be given to four groups of measures. First, support can be given for implementing local development strategies of local action groups with a view to achieving the objectives of one or more of the three other axes, as referred to in Article 62(1)(a). Next, support granted under the Leader axis must be for implementing cooperation projects for the objectives of the local development strategy. Third, support can be granted for running the local action group. And finally under axis 4, support can be given for acquiring skills and animating the territory as referred to in Article 59.

There is clearly the possibility of some overlap between the measures initiated under axis 4 and the measures that fall under the headings of the other axes. In such cases, Article 64 provides that if the operations under the local strategy correspond to the measures defined in the Regulation for the other axes, the relevant conditions for the other axes apply.[33]

[B] Strategic Guidelines, National Strategies and National Rural Development Programmes

The Rural Development Regulation only lays down the overall framework for what measures can be included in a rural development policy. Article 9(1) provides that the Council shall adopt Community strategic guidelines for rural development policy at Union level, setting the strategic priorities for rural development with a view to implementing each of the axes. This involves a choice between and a weighting of the various possible measures which are allowed under the Regulation, as well as striving for greater transparency about rural development policy.[34]

On this basis, in 2006 the Council laid down the framework for rural development policy in the Community strategic guidelines for rural development for 2007–2013.[35] Article 10 of the Rural Development Regulation allows for the Union's strategic

---

guidelines for the Community initiative for rural development (Leader+) (OJ C 139, 18 May 2000, 5), as amended by Commission communication amending the notice to the Member States of 14 Apr. 2000 laying down guidelines for the Community Initiative for rural development (Leader+) (OJ C 294, 4 Dec. 2003, 11).

33. Article 64 does not refer to the implementation of 'projects', which is the term otherwise used for axis 4, but refers to 'operations'. In Art. 2(e), an 'operation' is defined as 'a project, contract or arrangement, or other action selected according to criteria laid down for the rural development programme concerned and implemented by one or more beneficiaries allowing achievement of the objectives set out in Article 4.' See §5.02 above on Art. 4's definition of the objectives of support for rural districts.

34. See recital 8 of the Rural Development Regulation.

35. Council Decision 2006/144/EC on Community strategic guidelines for rural development (programming period 2007 to 2013) (OJ L 55, 25 Feb. 2006, 20). The strategic guidelines were adopted in 2006 pursuant to the legislative procedure in Art. 37 of the EC Treaty. On this procedure and the current legislative procedure in Art. 43(2) TFEU, see Ch. 3, §3.05.

guidelines to be revised during their period of validity in order to take account of major changes in Union priorities.[36]

Following the adoption of the Union strategic guidelines, according to Article 12 each Member States must prepare a national strategic plan. The priorities at Union level, especially in the strategic guidelines, form the basis for the national plans. Pursuant to Article 11(1), the plans must indicate the priorities of the actions of the EAFRD and of the Member States. The requirements for the content of the strategic plans are set out in Article 11(3), and include showing the consistency between the choices made in the national plan and the Union's strategic guidelines.[37] A persistent criticism of the rural development policy has been of its uneven implementation in the Member States. Much of the criticism has come from Union institutions.[38]

The Regulation does not expressly state that the Commission must approve each Member State's strategic plan, but it is assumed that the Commission will have a significant influence on the form of individual plans. In Article 11(1), it is stated that each Member State must 'submit' a national strategy plan. Article 12(1), second paragraph, says that the plan must be drawn up in 'close collaboration' with the Commission. Finally, in Article 12(2) it is stated that each Member State must send its national strategy plan to the Commission before submitting its rural development programmes.

The EAFRD acts through one or more rural development programmes in each Member State. According to Article 15(2), a Member State may either submit a single programme for its entire territory or a set of regional programmes. Article 15(1) states that the purpose of the programmes is to create a framework for implementing the strategy for rural development through a set of measures grouped together in accordance with the axes, for the achievement of which aid will be sought from the EAFRD.

The requirements for the content of individual programmes are set out in Article 16. Among other things, each programme must contain a justification of the priorities chosen having regard to the Community strategic guidelines and the national strategy plan, as well as the expected impact according to the ex ante evaluation. It must also contain information about the measures proposed for each axis, including specific verifiable objectives and indicators that allow the programme's progress, efficiency and effectiveness to be measured.

Pursuant to Article 18, the Commission assesses the programmes submitted by Member States on the basis of their consistency with the Union strategic guidelines, the national strategy plan and the Rural Development Regulation.

If the Commission is of the view that there is insufficient congruence between the programme and the sets of rules referred to, the Member State must revise its proposed programme. Individual rural development programmes are approved by the Rural

---

36. In 2011, the Commission submitted a report to the Council and the European Parliament on the implementation of national strategies and the Union guidelines for rural development (COM(2011) 450 final).
37. See Cardwell, 'Rural Development in the European Community: Charting a New Course?', Drake J. Agric. L 21 (39) (2008).
38. *Ibid.* 54–59. On the implementation of the rural development policy in France, see Bodiguel, Drake J. Agric. L 7 (2008).

Development Committee appointed pursuant to Article 90(2). Article 90(2) refers to the examination procedure laid down in Articles 4 and 7 of Regulation (EU) No 182/2011 laying down the rules and general principles concerning mechanisms for control by Member States of the Commission's exercise of implementing powers.[39] For some Member States this has been a burdensome process, e.g., Portugal's rural development programme was first approved in November 2008.[40]

---

39. Article 90(2) of the Rural Development Regulation refers to Arts 4 and 7 of Decision 1999/468/EC. This Decision was repealed by Regulation (EU) No 182/2011 and according to its Art. 13(1)(b) where a basic act refers to Art. 4 of Decision 1999/468/EC, the examination procedure referred to in Art. 5 of Regulation (EU) No 182/2011 applies.
40. See Commission Press Release of 19 Nov. 2008, IP/08/1734.

CHAPTER 6

The Single Payment Scheme

§6.01 THE SINGLE PAYMENT SCHEME AND ITS BASIS IN LEGISLATION

The single payment scheme is one of the elements of the legislation that is linked to the 2003 reform of the Union's agricultural policy.[1] The scheme was introduced by Council Regulation (EC) No 1782/2003 as a new form of income support for farmers. The rules for this scheme are now to be found in Council Regulation (EC) No 73/2009 establishing common rules for direct support schemes for farmers under the common agricultural policy and establishing certain support schemes for farmers. Article 1(b) states that one of the main purposes of the Regulation is to establish an income-support scheme for farmers which it refers to as the 'single payment scheme'. The most important provisions on the scheme are in Title III, Articles 33–72.

§6.02 THE CONDITIONS FOR THE PAYMENT OF AID PURSUANT TO THE SINGLE PAYMENT SCHEME, INCLUDING THE REQUIREMENT FOR AGRICULTURAL ACTIVITY

Regulation (EC) No 73/2009 lays down four main conditions for a farmer to receive support from the single payment scheme. First, he must be a 'farmer' within the meaning of the Regulation. According to Article 33(1) of the Regulation, support under the single payment scheme is only to be available to farmers, and 'farmers' are defined in Article 2(a) of the Regulation as a natural or legal person who exercises an agricultural activity.[2] The provision does not make it a requirement that the farmer should own the holding, only that he should exercise an agricultural activity, so that both owners and tenants of agricultural holdings are covered by the single payment

---

1. For a general review of the 2003 reforms, see Ch. 1, §1.01.
2. On the concept of a 'farmer', see Ch. 2, §2.03. Art. 2(a) of the Regulation also requires that the farmer's holding is situated within Union territory.

scheme. In connection with the single payment scheme, according to Article 2(c) 'agricultural activity' means the production, rearing or growing of agricultural products including harvesting, milking, breeding animals and keeping animals for farming purposes, or maintaining the land in good agricultural and environmental condition.[3] Forestry is not an agricultural activity and is thus not covered by the single payment scheme.[4]

The second main condition for receiving aid under the single payment scheme is that the farmer should have agricultural land at his disposal that is eligible for support; see Article 34 of Regulation (EC) No 73/2009. In the Regulation, this land is referred to as 'eligible hectares', and this term is defined in Article 34(2).[5]

The third main condition for receiving aid under the single payment scheme, pursuant to Article 33(1), is that farmers should hold payment entitlements. Article 34(1) links together the requirements for farmers to have agricultural land at their disposal that is eligible for support and to hold payment entitlements. In order to receive payments under this provision, a farmer must activate a payment entitlement in respect of eligible hectares.[6]

The fourth and last condition that farmers must satisfy in order to be covered by the single payment scheme is that they must respect the statutory management (cross compliance) requirements; see Article 4 of Regulation (EC) No 73/2009.[7]

§6.03   ELIGIBLE HECTARES

Article 34(2) of Regulation (EC) No 73/2009 defines and divides eligible hectares or areas into two main categories. The first main category is defined in Article 34(2)(a) as any agricultural area of the holding, and any area planted with short rotation coppice. The general rule of the scheme is that eligible hectares or areas are the agricultural areas of a holding. In Article 2(h), an 'agricultural area' is defined as any area taken up by arable land, permanent pasture or permanent crops. This definition of agricultural area means that both cultivated and uncultivated areas are regarded as eligible areas. As for areas planted with short rotation coppice, according to Article 34(2)(a), it is a condition that areas are used for an agricultural activity or, if an area is also used for non-agricultural activities, that it is predominantly used for agricultural activities.[8]

The second main category of areas eligible to receive support consists of any area which gave a payment entitlement under the single payment scheme or the single area payment scheme in 2008 but which, for certain reasons, is no longer considered

---

3.   Article 6 of the Regulation defines in more detail what is meant by good agricultural and environmental condition.
4.   On forestry and the rules of Union agricultural law, see Ch. 2, §2.01.
5.   See §6.03 on the term 'eligible hectares'.
6.   On the term 'eligible hectares' see §6.03, and on the activation of payment entitlements in respect of eligible hectares see §6.05.
7.   On cross compliance requirements, see Ch. 7.
8.   In Case C-61/09 *Landkreis Bad Dürkheim* [2010] ECR I-9763, it was established that it is not necessary, for an agricultural area to be considered as allocated to the farmer's holding, that it be at his disposal against payment on the basis of a lease or another similar type of contract to let,

'eligible' under Article 34(2)(a). According to Article 34(2)(b)(i), these areas are nevertheless regarded as eligible areas if the reason for the area in question no longer being eligible is due to the implementation of Directive 2009/147/EC on the conservation of wild birds, Council Directive 92/43/EEC on the conservation of natural habitats and of wild fauna and flora, and Directive 2000/60/EC establishing a framework for Community action in the field of water policy.[9] Similarly, pursuant to Article 34(2)(b)(ii), areas are eligible for support even if they do not fulfil the requirements of Article 34(2)(a) if the areas in question have been afforested pursuant to Article 31 of Council Regulation (EC) No 1257/1999 or to Article 43 of Regulation (EC) No 1698/2005 or under a national scheme the conditions of which comply with Article 43(1), (2) and (3) of the latter Regulation.[10] Finally, areas that do not fulfil the conditions in Article 34(2)(a) are regarded as being entitled to payments if the areas are set aside pursuant to Articles 22, 23 and 24 of Regulation (EC) No 1257/1999 or to Article 39 of Regulation (EC) No 1698/2005.[11]

The conditions that apply to an area in order for it to be considered eligible for support must be complied with throughout the calendar year, according to Article 34(2). This provision makes an exception in the event of force majeure or exceptional circumstances.[12]

§6.04 OBTAINING PAYMENT ENTITLEMENTS

Pursuant to Article 33(1) of Regulation (EC) No 73/2009, a farmer can obtain payment entitlements in several ways. First and foremost, a farmer can hold payment entitlements which he has obtained in accordance with Regulation (EC) No 1782/2003 when it was in force (the predecessor to Regulation (EC) No 73/2009; see section §6.01 above). These payment entitlements were allocated to farmers in 2005 upon the introduction of the single payment scheme. Article 43(1) of Regulation (EC) No 1782/2003 provided that each farmer received payment entitlements in respect of each hectare at their disposal.[13]

In addition to the initial allocation of payment entitlements on the basis of Regulation (EC) No 1782/2003, a farmer can obtain or be allocated payment entitlements in various ways, as laid down in Article 33(1)(b)(i) to (iv) of Regulation (EC) No 73/2009. These possibilities of acquiring payment entitlements are discussed in sections §6.04[A] to §6.04[D] below.

---

9. The actual wording of Art. 34(2)(b)(i) refers to Council Directive 79/409/EEC and not to Directive 2009/147/EC as shown in the text above. Directive 79/409/EEC was repealed and replaced by the provisions in Directive 2009/147/EC. Article 18 of Directive 2009/147/EC provides that references to Directive 79/409/EEC in Union legislative acts must be construed as references to Directive 2009/147/EC, and read in conjunction with the correlation table in Annex VII to the Directive.
10. Article 43 of Regulation (EC) No 1698/2005 (the Rural Development Regulation) concerns the first afforestation of agricultural areas.
11. Article 34(2)(b)(iii). Article 39 of the Rural Development Regulation concerns support for carrying on environmentally friendly agriculture.
12. On force majeure, see Ch. 4, §4.06[C].
13. The calculation of the relevant number of hectares was relatively complex; see Art. 43 of Regulation (EC) No 1782/2003.

[A]   Obtaining Payment Entitlements by Transfer: Article 33(1)(b)(i)

Article 33(1) in Regulation (EC) No 73/2009 sets out several independent ways in which a farmer can acquire payment entitlements. As the most important of these, Article 33((1)(b)(i) provides that a farmer can obtain payment entitlements by transfer. The rules for transfers are specified in Article 43. According to Article 43(2), payment entitlements may be transferred by sale or by any other definitive transfer, as well as by lease or 'similar types of transactions'.[14] In the sense in which transfer is used in the Regulation, transfers of payment entitlements can be made definitively in the form of a sale or temporarily by a lease or a similar transaction. Where a definitive transfer is made, Article 43(2) allows payment entitlements to be transferred with or without land.[15] In contrast, under Article 43(2), second sentence, a lease or a similar transaction of payment entitlements is only allowed if the payment entitlements are transferred together with an equivalent number of eligible hectares.[16]

Other than in the case of the transfer by actual or anticipated inheritance pursuant to Article 43(1), payment entitlements can only be definitively or temporarily transferred to a farmer who is established in the same Member State as the transferor. However, even in the case of transfer by actual or anticipated inheritance, payment entitlements may be used only in the Member State where the payment entitlements were established.[17]

[B]   Obtaining Payment Entitlements from the National Reserve: Article 33(1)(b)(ii)

The second method by which a farmer can obtain payment entitlements is from the national reserve. According to Article 41(1) of Regulation (EC) No 73/2009, each Member State operates with a national reserve which can be characterized as a special pool of payment entitlements which can be allocated to farmers upon application. The national reserve of an individual Member State consists of the difference between the national ceilings for the total financial value of payment entitlements in the Member State in question and the total value of all allocated payment entitlements, together

---

14. Article 23(1) of Regulation (EC) No 73/2009, on the reduction of, or exclusion from, support payments in the event of failure to fulfil cross compliance requirements states that: 'For the purpose of this paragraph, "transfer" shall mean any type of transaction whereby the agricultural land ceases to be at the disposal of the transferor'. On 'definitive transfer', see Case C-434/08 *Arnold and Johan Harms* [2010] ECR I-4431.
15. In connection with the sale of payment entitlements with or without land, Art. 43(3) gives individual Member States the right to lay down national rules under which part of the payment entitlements sold must revert to the national reserve, or their unit values may be reduced in favour of the national reserve.
16. In Case C-470/08 *Kornelis van Dijk* [2010] I-603, it was repeated (under Regulation 1782/2003) that Community Law does not require a lessee, on the expiry of the lease, to deliver to the lessor not only the leased land, but also the payment entitlements accumulated thereon or relating thereto, or to pay him compensation.
17. According to Art. 43(1), third sentence, a Member State may decide that payment entitlements may be transferred or used only within one and the same region.

with a special ceiling for payments for certain agricultural products and a special ceiling for specific support.[18]

Article 41(2) to (6) sets out the purposes to which individual Member States may put the payment entitlements in their national reserves. These provisions thus set limits to how the rights can be acquired. It is a general condition of the provisions that a Member State must allocate the payment entitlements according to objective criteria and in such ways as to ensure equal treatment of farmers, and so as to avoid distortion of the market or of competition. Article 41(2) authorizes Member States to allocate payment entitlements to farmers who are starting their agricultural activities. Article 41(3) authorizes Member States to allocate payment entitlements to farmers in areas subject to restructuring and/or development programmes to ensure against land being abandoned and/or to compensate farmers for specific disadvantages in those areas.[19] Under Article 41(4) Member States may use their national reserve to allocate payment entitlements to farmers placed in a 'special situation'. According to the provision, this 'special situation' is to be defined by the Commission and this has been done in Regulation (EC) No 1120/2009. Article 19 in Regulation (EC) No 1120/2009 states that for the purposes of Article 41(4) of Regulation (EC) No 73/2009, 'farmers in a special situation' shall mean the farmers referred to in Articles 20–23 of Regulation No 1120/2009. Finally, there is a special rule in Article 41(6) for Member States that use Article 59 of Regulation (EC) No 73/2009 on the allocation of payment entitlements in new Member States that have used general area payment schemes, or which use Article 63 on the integration of coupled support into the single payment scheme. Article 41(6) gives such Member States the right, under certain conditions, to provide that some or all of the payment entitlements or of the increase in the value of payment entitlements that would be allocated to a farmer shall revert to the national reserve if the allocation or increase would lead to a windfall profit for the farmer in question. The situations in which a transfer to the national reserve may be appropriate are in the event of a sale or grant or expiry of all or part of a lease of a holding or of premium rights.

[C] Obtaining Payment Entitlements Pursuant to Annex IX: Article 33(1)(b)(iii)

Article 33(1)(b)(iii) of Regulation (EC) No 73/2009 provides that farmers may obtain payment entitlements pursuant to the rules in Annex IX to the Regulation. The Annex concerns three groups of payment entitlements; its Point A deals with fruit and vegetables, ware potatoes and nurseries; Point B is concerned with grubbing-up schemes for vines; and Point C is concerned with support programmes in the wine sector.

---

18. The ceilings for the total financial value of payment entitlements of the Member States are set out in Annex VII of the Regulation. For the special ceilings relating to sheep and goat meat, beef and transitional payments for fruit and vegetables, see Art. 51(2); and for the special ceiling for specific support, see Art. 69(3).
19. It is a condition for Member States being allowed to use the national reserve under this heading that they do not apply Art. 68(1)(c). Under Art. 68, Member States may give support to farmers in areas that are subject to restructuring and/or development programmes.

Under Point A(1), second paragraph, farmers receive a payment entitlement per-hectare in respect of fruit and vegetables or ware potatoes.[20] The number of hectares is calculated by dividing the reference amount specified in Point A(2) by the 'objective' number of hectares as calculated in accordance with the rules in Point A(3).

Point A(2) leaves it to the Member States to establish the rules for how the reference amount is to be calculated. However, each Member State must use objective and non-discriminatory criteria, and examples are given of how the reference amount can be determined. The examples use either the market support received by the farmer, or the area used for production, or the quantity of fruit and vegetables or ware potatoes produced. Market support, the area used, and the quantity produced can also be used for determining the reference amount for nurseries. Point A(2) give express authority for Member States to use differing criteria for different fruit and vegetables, ware potatoes and nurseries, if this is objectively justified.

In the same way that Point A(2) leaves it to the Member States to determine how reference amounts are to be calculated, Point A(3) gives the Member States the authority to lay down the rules for calculating the 'objective' number of hectares. The Member States must use objective and non-discriminatory criteria, and an example is given of such a criterion in the area used for the production of fruit and vegetables, ware potatoes and nurseries; see Point A(2)(1)(b).

The Annex lays down provisions on force majeure in Point A(4) to (6).[21]

Under the provisions in Point B, farmers who participate in the grubbing-up scheme for vines laid down in Council Regulation (EC) No 479/2008 on the common organization of the market in wine are allocated payment entitlements after the grubbing-up corresponding to the number of hectares for which they receive a grubbing-up premium. Point B contains special rules for determining the unit value of payment entitlements.

Where a Member State has granted support pursuant to the rules for the common organization of the market in wine pursuant to Regulation (EC) No 479/2008, under Point C of Annex IX farmers receive one unit of payment entitlement per-hectare. The number of hectares is calculated by a principle corresponding to the calculation of hectares of fruit and vegetables etc. under Point A.

[D]   Obtaining Payment Entitlements in Connection with Certain Special Arrangements: Article 33(1)(b)(iv)

This category of the ways in which a farmer can obtain payment entitlements is a collection of various special arrangements. The category is defined in Article 33(1)(b)(iv) by reference to Article 47(2), Article 59, Article 64(2), third paragraph, Article 65 and Article 68(4)(c).

---

20.   According to Part A(1), first paragraph, the definitions of 'fruit and vegetables' are as given in Art. 1(1)(i) and (j) of Regulation (EC) No 1234/2007, and the definition of 'ware potatoes' is as given in CN code 0701.
21.   On force majeure, see Ch. 4, §4.06[C].

Article 47(2) concerns Member States that have decided to use a special authority for regionalization of the single payment scheme.[22] Article 59 contains rules that apply to new Member States that have used a general area payment scheme.[23] Article 64(2) and Article 65 are linked to rules that apply to all Member States, and concern the integration of coupled support in the single payment scheme. Finally, Article 68(4)(c) concerns the provision of special support in areas where there are restructuring and development programmes. According to this provision, this takes the form of an increase in the unit value and/or the number of the farmer's existing payment entitlements.

§6.05   THE USE OF ELIGIBLE HECTARES

The requirements that a farmer should have at his disposal areas that are eligible for support and payment entitlements are linked together in Article 35(1) of Regulation (EC) No 73/2009. According to Article 35(1), the farmer must declare the parcels corresponding to the eligible hectares accompanying any payment entitlement. It is up to the Member State to fix the date when these parcels must be at the farmer's disposal, but this date must be no later than the date fixed by that Member State for amending an aid application. However, the date for amending an aid application can in fact be extended, and Article 35(2) allows the authorities of a Member State to authorize a farmer to modify his declaration in duly justified circumstances. However, the grant of such authority is conditional on the farmer adhering to the number of hectares corresponding to his payment entitlements and the conditions for granting the single payment for the area concerned.

The granting of support payments is subject to the condition that farmers fulfil the cross compliance requirements.[24]

---

22.   See Arts 46-50 of Regulation (EC) No 73/2009.
23.   See Arts 55-62 of Regulation (EC) No 73/2009.
24.   On cross compliance requirements, see Ch. 7.

CHAPTER 7
Cross Compliance Rules

---

§7.01 INTRODUCTION TO THE CROSS COMPLIANCE RULES AND THEIR LEGISLATIVE BASIS

'Cross compliance' is a concept of Union agricultural law which is of major practical importance. The term refers to systematic requirements applicable to farmers and it means that farmers who receive direct support or subsidies under certain arrangements under a rural development programme must comply with a number of requirements relating to the environment, animal welfare and good agricultural practice in order to be paid the full amount of the support or subsidy. The support or subsidy can be reduced or denied in full if these requirements are not fully satisfied. It is a condition of cross compliance that the requirements must be complied with for the whole agricultural holding if the agricultural aid is to be paid in full. In this context an agricultural holding includes all the production units operated by the farmer in the same Member State.[1]

The cross compliance rules were originally introduced in Regulation (EC) No 1782/2003 with effect from 2005. This Regulation was repealed in 2009 and today the legislative basis for cross compliance requirements is Council Regulation (EC) No 73/2009 establishing common rules for direct support schemes for farmers under the common agricultural policy and establishing certain support schemes for farmers. The requirement for cross compliance not only applies to direct support, but also to certain parts of rural development aid. While the scope of Regulation (EC) No 73/2009 does not itself extend to rural development aid, it is clear from Articles 50a and 51 of Council Regulation (EC) No 1698/2005 on support for rural development by the EAFRD (the 'Rural Development Regulation') that the cross compliance requirements also apply to support for sustainable use of agricultural areas and forestry areas.[2]

---

1. See Art. 23(1) of Council Regulation (EC) No 73/2009.
2. See Art. 36(a)(i) to (v) and Art. 36(b)(i), (iv) and (v) of Regulation (EC) No 1698/2005, on axis 2 of rural development policy. On rural development policy and on its axes, see Ch. 5 above.

In Regulation (EC) No 1122/2009, the Commission has laid down detailed rules for the implementation of Council Regulation (EC) No 73/2009 as regards cross compliance (hereinafter the 'Implementation Regulation').[3]

§7.02 THE PURPOSE OF THE CROSS COMPLIANCE REQUIREMENT

There is no statement in Union legislation of the overall purpose of cross compliance requirements.[4] However, it must be assumed that the primary aims of cross compliance requirements are to promote sustainable agriculture and rural development.[5] In the preamble to Regulation (EC) No 73/2009 it is stated that Regulation (EC) No 1782/2003 established the principle that farmers who do not comply with certain requirements in the areas of public, animal and plant health, the environment and animal welfare are subject to reductions of or exclusion from direct support. It is stated that the cross compliance system forms an integral part of Union support for farmers.[6] The linking together of the requirements in the various support schemes, each of which has the overall aim of contributing to agriculture and rural development, must be assumed to have the same purpose.

Another purpose of the cross compliance requirements is to link together requirements in different support measures so that payments of support or subsidies pursuant to a specific scheme are conditional on fulfilment of requirements linked to other support or subsidy schemes. A support or subsidy is referred to in the legislation as 'direct payment', which means a payment granted directly to farmers under a support scheme listed in Annex I to Regulation (EC) No 73/2009.[7]

Arts 50a and 51 of the Rural Development Regulation originally referred to Regulation (EC) No 1782/2003, but Art. 1(9)-(13) of Regulation (EC) No 74/2009 amended this to a reference to Regulation (EC) No 73/2009 and to the content of its provisions. According to Art. 146(2) of Regulation (EC) No 73/2009, references in other legislation to Regulation (EC) No 1782/2003 are to be construed as reference to Regulation (EC) No 73/2009. Annex XVIII to Regulation (EC) No 73/2009 contains a correlation table of the rules in Regulation (EC) No 1782/2003 and Regulation (EC) No 73/2009.

3. Commission Regulation (EC) No 1122/2009 repeals and replaces Commission Regulation (EC) No 796/2004.
4. The European Court of Auditors has identified three overall goals of the cross compliance requirements in recitals 2, 3 and 4 of Council Regulation (EC) No 1782/2003, but it criticized the lack of specific and measurable targets; see European Court of Auditors, *Is Cross Compliance an Effective Policy?* 12–13, Special Report No 8/2008.
5. Recital 8 in Regulation (EC) No 73/2009 refers to the fact that Regulation (EC) No 1782/2003 introduced a system of compulsory progressive reduction of direct payments to farmers (modulation), and that this was intended to achieve a better balance between policy tools designed to promote sustainable agriculture and those designed to promote rural development. On the aims of the Rural Development Regulation, see Ch. 5, §5.02[A].
6. See recital 3 in Regulation (EC) No 73/2009. In the Rural Development Regulation there is only an express reference to the cross compliance requirement in recital 15. On the specific purposes otherwise pursued by Art. 51 of the Rural Development Regulation, see §7.03[B] below.
7. See Art. 2(d) of Regulation (EC) No 73/2009. The Rural Development Regulation does not contain any definition of 'direct payment', but it must be assumed to have the same meaning both under Regulation (EC) No 73/2009 and under the Rural Development Regulation.

§7.03 THE OBLIGATIONS OF RECIPIENTS OF SUPPORT UNDER THE CROSS COMPLIANCE REQUIREMENTS

The cross compliance requirements are set out in Articles 4, 5 and 6 of Regulation (EC) No 73/2009 and Articles 50a and 51 of the Rural Development Regulation. Even though there are only five relatively short provisions, the requirements are in fact comprehensive as these provisions to a large extent refer to a large number of requirements that are laid down in other Union legislative acts. Furthermore, in many cases the Union legislation depends on further and more detailed requirements being laid down in national law. Thus many of the cross compliance requirements are based on Union legislation in the form of directives which require the Member States to establish or to ensure the existence of a given legal situation. In these cases, it is not the farmer at whom the Union legislation is addressed, as the specific cross compliance provisions are to be incorporated in national legislation giving effect to Union law. While such cross compliance requirements are authorized by Union rules, their specific content is to be found in the national rules of the Member State. A full overview of all cross compliance requirements is thus only possible by taking together the rules of Union law and the national rules of the Member State in question.

[A]   Cross Compliance Requirements in Connection with Direct Support

Article 4(1) of Regulation (EC) No 73/2009 lays down what the heading of the Article itself refers to as 'Main requirements' of cross compliance: 'A farmer receiving direct payments shall respect the statutory management requirements listed in Annex II and the good agricultural and environmental condition referred to in Article 6'.[8]

According to Article 5, statutory management requirements listed in Annex II to the Regulation relate to three areas, and Annex II is thus divided into three parts. However, the more detailed division of the Annex makes it clear that two further groups of requirements are also covered by the Article 5 areas.[9] The specification of the statutory management requirements in Annex II to the Regulation is made by reference to specific provisions in specific regulations and directives. In summary, the cross compliance requirements connected with direct support, together with the requirements in Article 6, cover six areas:

- The environment.
- Identification and registration of animals.
- Public, animal and plant health.
- Notification of diseases.
- Animal welfare.
- The condition of agricultural land.

---

8.   Article 2(31) of the Implementation Regulation defines 'cross-compliance' as compliance with the statutory management requirements and the good agricultural and environmental condition in accordance with Arts 5 and 6 of Regulation (EC) No 73/2009.
9.   On the background to the cross compliance requirements, see Daniele Bianchi, *La politique agricole commune (PAC)* 280-284 (Bruylant, 2nd Ed., 2012).

In all of these areas, some of the requirements only concern specific sectors of agriculture, but both for national authorities and in particular for farmers, these are comprehensive and complex requirements.[10] Where, in the following, in the review of Union legislation there is a reference to a cross compliance requirement, this should often be understood as a reference to a group of cross compliance requirements. This is because, as stated above, Union legislation often takes the form of a directive which lays down some more or less general requirements which must be implemented as several more detailed requirements in national law.

According to Article 4(1), second paragraph, the main requirements apply only to the agricultural activity of the farmer or to the agricultural area of the holding concerned.

Presumably as a form of guarantee of legal certainty, Article 4(2) provides that national authorities must provide the farmer with the list of statutory management requirements and the good agricultural and environmental condition to be respected.

[1]   Environmental Requirements

In the area of the environment, there are statutory management requirements in five different legislative acts. The overall aim of the environmental requirements is that they should help protect nature, the environment and the landscape so there can be development on a sustainable basis.

For the first requirement, Annex II(A)(1) to Regulation (EC) No 73/2009 refers to several provisions in Council Directive 79/409/EEC.[11] According to its Article 1, the Directive concerns the conservation of all species of naturally occurring birds in the wild state in the European territory. Article 3(1) requires the Member States to take the requisite measures to preserve, maintain or re-establish a sufficient diversity and area of habitats for all the protected species of birds. Article 3(2)(b) states that the preservation, maintenance and re-establishment of biotopes and habitats include primarily the upkeep and management of habitats inside and outside the protected zones, in accordance with the ecological needs. There is an Annex to the Directive with a list of the bird species that according to Article 4(1) of the Directive are the subject of special conservation measures concerning their habitat in order to ensure their survival

---

10. In connection with the earlier Regulation (EC) No 1782/2003, which established the cross compliance arrangements, the European Court of Auditors pointed out the difficulties associated with the fact that farmers had to comply with so many and such complex statutory management requirements. Among other things the Court of Auditors pointed out that the cross compliance arrangements then in place included a long list with requirements that had to be monitored and that, for example, in 2007 Dutch inspectors used a checklist which referred to 172 different rules which had to be complied with and monitored; see European Court of Auditors, *Is Cross Compliance an Effective Policy?* 18 Special Report No 8/2008.

11. While Annex II(A)(1) refers to Council Directive 79/409/EEC, this Directive has been repealed and replaced by provisions in Directive 2009/147/EC on the conservation of wild birds. According to Art. 18 of Directive 2009/147/EC, references in Union legislative acts to Directive 79/409/EEC are to be construed as references Directive 2009/147/EC, and must be read in accordance with the correlation table in Annex VII. The references to Art. 3(1), Art. 3(2)(b), Art. 4(10, (2) and (4), and Art. 5(b) and (d) in Directive 79/409/EEC should be construed as references to the same article numbers in Directive 2009/147/EC.

and reproduction in their area of distribution. The provision obliges the Member States to classify territories that are suitable in number and size as special protection areas for the conservation of these species. Article 4(2) extends the obligation of Member States, under certain conditions, to take special conservation measures that cover regularly returning migratory bird species that are not listed in Annex 1. Article 4(4) requires the Member States, under certain circumstances, to take appropriate steps to avoid pollution or deterioration of habitats or any disturbances affecting the birds referred to Article 4(1) and (2). Also, under Article 5(b) and (d), the Member States are required to take the requisite measures to establish a general system of protection for all the species of birds referred to in Article 1 of the Directive. In connection with the cross compliance requirements, this includes prohibiting the deliberate destruction of, or damage to, nests and eggs or removal of nests. There is also a special cross compliance requirement prohibiting the deliberate disturbance of birds particularly during the breeding season, in so far as disturbance would be significant to the objectives of the Directive.

Annex II(A)(2) to Regulation (EC) No 73/2009 refers to Articles 4 and 5 of Council Directive 80/68/EEC on the protection of groundwater against pollution caused by certain dangerous substances.[12] Article 3(a) of the Directive requires the Member States among other things to take the necessary steps to prevent the introduction into groundwater of substances that are listed in an annex to the directive as list I. As a general rule, Article 4 of the Directive obliges the Member States to prohibit all direct discharge of substances in list I. The Member States are under the same obligation though within broader frames when it comes to the introduction into groundwater of substances in list II in the annex.

Annex II(A)(3) to the Regulation refers to Article 3 of Council Directive 86/278/EEC on the protection of the environment, and in particular of the soil, when sewage sludge is used in agriculture. The requirement relates to the use of residual sludge from sewage plants treating domestic or urban waste waters and residual sludge from septic tanks. Sewage sludge from domestic or urban waste waters may only be used in agriculture in accordance with the requirements laid down in the Directive. As for residual sludge from septic tanks and other similar installations for the treatment of sewage, this can in principle only be used in agriculture in accordance with the conditions imposed by the Member State in question. According to Article 12 of the Directive, where conditions so demand, Member States may take more stringent measures than those provided for in the Directive.

The fourth cross compliance requirement is referred to in Annex II(A)(4) to the Regulation, which refers to Articles 4 and 5 of Council Directive 91/676/EEC on the protection of waters against pollution caused by nitrates from agricultural sources. Article 4 of this Directive requires the Member States to establish a code or codes of good agricultural practice, to be implemented by farmers on a voluntary basis. The code should contain provisions covering at least the items mentioned in Annex II(A) to

---

12. Directive 80/68/EEC has been repealed and replaced by the rules in Directive EC/2000/60 with effect from 21 Dec. 2013; see Art. 22(2) of the latter Directive.

the Directive. These items concern measures to reduce nitrate pollution from agriculture. Among other things, these cover periods when the land application of fertilizer is inappropriate; the conditions for land application of fertilizer near water courses; and the capacity and construction of storage vessels for livestock manures. As stated, in isolation compliance with such a code is voluntary for farmers, but if a farmer does not comply with a code they will not be cross compliant, and this will have consequences for their rights to receive support payments.[13] Article 5 requires the Member States to draw up action programmes for designated vulnerable zones. Such action programmes must consist of a number of mandatory measures, including special measures referred to in Annex III to the Directive. These measures deal with periods when the land application of certain types of fertilizer is prohibited.

In Annex II(A)(5) to Regulation (EC) No 73/2009, the fifth and final cross compliance requirement is linked to Articles 6 and 13(1)(a) of Council Directive 92/43/EEC on the conservation of natural habitats and of wild fauna and flora. Article 6 of this Directive requires the Member States to establish the necessary measures for special conservation areas. These areas are defined in Article 1(l) as sites of Union importance designated by the Member States through a statutory, administrative and/or contractual act where the necessary conservation measures are applied for the maintenance or restoration, at a favourable conservation status, of the natural habitats and/or the populations of the species for which the site is designated. Similarly, Article 13(1)(a) requires Member States to take the necessary measures to establish a system of strict protection for the plant species listed in Annex IV(b). These measures should prohibit the deliberate picking, collecting, cutting, uprooting or destruction of such plants in their natural range in the wild. The prohibition applies to all stages of the biological cycle of the plants to which the Article applies.

[2]   *Requirements for the Identification and Registration of Animals*

In addition to containing categories of cross compliance requirements relating to the environment, Annex II(A)(6) to (8) to Regulation (EC) No 73/2009 contain categories of requirements for the identification and registration of animals.

Annex II(A)(6) refers to Articles 3, 4 and 5 of Council Directive 2008/71/EC on identification and registration of pigs. Articles 3 and 4 require Member States to ensure that their competent authority has an up-to-date list of all the holdings which keep animals covered by the Directive and situated on its territory and that any keeper so listed keeps a register stating the number of animals present on the holding. Article 5 lays down requirements for identification marking of animals. According to Article 5(2), animals must be marked with an ear tag or tattoo as soon as possible, and in any case before they leave the holding, making it possible to determine the holding from which they come and enabling reference to be made to any accompanying document which must mention such ear tag or tattoo and to the list referred to in Article 3(1)(a).

---

13.   See §7.04[C] below on reductions of or exclusions from support payments.

The requirements for marking and identification of the animals on a holding are further specified in Annex II(A)(7)'s reference to Articles 4 and 7 of Regulation (EC) No 1760/2000 establishing a system for the identification and registration of bovine animals and regarding the labelling of beef and beef products.

There are special requirements for the registration and marking of sheep and goats. Annex II(A)(8) refers to Articles 3, 4 and 5 of Council Regulation (EC) No 21/2004. Article 1 of the Regulation provides that each Member State must establish a system for the identification and registration of sheep and goats. According to Article 3, such a system must comprise the means of identification to identify each animal; up-to-date registers kept on each holding; movement documents; and a central register or computer database. The individual elements of the system are specified in Articles 4 and 5.

[3] Requirements Relating to Public, Animal and Plant Health

The statutory management requirements, i.e., the cross compliance requirements in the area of public, animal and plant health are laid down in selected provisions in seven different legislative acts.[14]

The first requirement concerns plant protection products. Annex II(B)(9) to Regulation (EC) No 73/2009 refers to Article 55 of Regulation (EC) No 1107/2009 on the placing of plant protection products on the market.[15] Article 55 states that plant protection products must be used properly, and proper use means the application of the principles of good plant protection practice and compliance with the conditions established in accordance with Article 31 (of Regulation (EC) No 1107/2009) and specified on the labelling. Proper use also means that use should be in accordance with the provisions of Directive 2009/128/EC and, in particular, with general principles of integrated pest management, as referred to in Article 14 of Annex III to that Directive.[16]

The second category under the heading of public, animal and plant health concerns the prohibition of the use of certain substances in animal husbandry. Annex II(B)(10) to Regulation (EC) No 73/2009 refers to Article 3(a), (b), (d) and (e) and Articles 4, 5 and 7 of Council Directive 96/22/EC on the prohibition on the use in stockfarming of certain substances having a hormonal or thyrostatic action and of beta-agonists. There is a general prohibition in Article 3, while Articles 4, 5 and 7 lay down some exceptions to the general prohibition. Article 3 states that Member States

---

14. See the legislative acts referred to in Annex II(B)(9) to (15) to Regulation (EC) No 73/2009.
15. The actually wording of Annex II(B)(9) refers to Art. 3 of Council Directive 91/414/EEC on the placing of plant protection products on the market. However, this Directive was largely repealed by Regulation (EC) No 1107/2009; see Arts 83 and 84 of the Regulation. According to Art. 83, second paragraph, references in other Union legislation, 'such as Regulation (EC) No 1782/2003' (i.e., the former Regulation on cross compliance), to Art. 3 of Directive 91/414/EEC are to be construed as references to Art. 55 of Regulation (EC) No 1107/2009. Since Regulation (EC) No 1782/2003's reference to Art. 3 in Directive 91/414/EEC was carried forward unchanged in Regulation (EC) No 73/2009, Art. 83, second paragraph, must be understood as meaning that Annex II(B)(9) to Regulation (EC) No 73/2009 now refers to Art. 55 of Regulation (EC) No 1107/2009.
16. Article 55 refers to the requirements in Art. 14 and Annex III to Directive 2009/128/EC.

must prohibit the use of substances listed in Annex II, and provisionally prohibit the use of substances listed in Annex III. Annex II lists, for example: thyrostatic substances, stilbenes, stilbene derivatives, their salts and esters, oestradiol 17β, and beta-agonists.[17] The provisionally prohibited substances in Annex III include: substances having oestrogenic (other than oestradiol 17β and its ester-like derivatives), andro-genic or gestagenic action. In connection with the cross compliance requirements, there is a prohibition of the administration of the named substances to domestic animals by farmers, and of the keeping of such animals on a farm, the placing on the market or slaughter for human consumption of farm animals or of aquaculture animals which contain such substances. The exceptions to the general prohibition in Articles 4, 5 and 7 relate primarily to the use of such substances in connection with a veterinary surgeon's treatment of animals.

Annex II(B)(11) to Regulation (EC) No 73/2009 lays down a third requirement regarding compliance with food law. The requirement refers to Articles 14 and 15, Article 17(1) and Articles 18, 19 and 20 of Regulation (EC) No 178/2002 laying down the general principles and requirements of food law. Articles 14 and 15 lay down detailed requirements for food safety and feed safety. As a general rule, Article 14(1) and (2) state that food may not be placed on the market if it is unsafe or dangerous. Food is regarded as dangerous if it is considered injurious to health or not fit for human consumption. Article 15(1) and (2) state that feed may not be placed on the market or fed to any food-producing animal if the feed is dangerous. Feed is considered to be dangerous if it is regarded as having an adverse effect on human or animal health or if it would make the food derived from food-producing animals unsafe for human consumption. Article 17(1) and Articles 18, 19 and 20 lay down a number of detailed requirements for individual operators of food and feed undertakings. According to Article 17(1), food and feed business operators must ensure that, at all stages of production, processing and distribution within the businesses under their control, the foods or feeds satisfy the requirements of food law which are relevant to their activities and must control that such requirements are met. Article 18 specifies a special requirement for the traceability of food. Food and feed business operators must be able to identify from whom they have been supplied with a food, a feed, a food-producing animal, or any substance intended to be, or expected to be, incorporated into a food or feed. For this purpose, such operators must have in place systems and procedures to enable this information to be made available to the competent authorities on demand. The requirement for traceability also applies to the undertakings which the food and feed business operators supply. Information about those to whom the food or feed has been supplied must also be made available to the competent authorities on demand. Articles 19 and 20 specify the responsibilities of food and feed business operators for food and feed safety.

The fourth category relates to combating certain forms of very serious diseases such as certain transmissible spongiform encephalopathies (TSE) (also known as Creutzfeldt-Jakob disease or BSE). Annex II(B)(12) to Regulation (EC) No 73/2009

---

17.   There are certain exceptions to the prohibition of beta-agonists.

refers to Articles 7, 11, 12, 13 and 15 of Regulation (EC) No 999/2001. This Regulation is aimed at preventing and eradicating these kinds of diseases. Article 7(1) lays down an absolute prohibition of feeding protein derived from mammals to ruminants. Article 7(2) extends the prohibition to animals other than ruminants to the extent laid down in detailed rules in Annex IV to the Regulation. Articles 11 and 12 lay down rules for how to deal with cases where an animal is suspected of being infected with TSE while Article 13 deals with cases where it is established that an animal is infected. As part of the campaign against TSE, Article 15 lays down special rules on the export of live animals, semen, embryos and ova.

[4] Requirements for Notification of Diseases

Annex II(B)(13) to Regulation (EC) No 73/2009 refers to Article 3(1)(a) of Council Directive 2003/85/EC.[18] This provision requires the Member States to ensure that foot-and-mouth disease is listed by a competent authority as a compulsorily notifiable disease. As with other provisions of directives, this is not binding on individual farmers, but the reference to the provision as a cross compliance requirement means that farmers are bound to comply with national rules on the notification of foot-and-mouth disease which implement the Directive in national law.[19]

Annex II(B)(14) to the Regulation refers to Article 3 of Council Directive 92/119/EEC. This provision requires Member States to ensure that it is compulsory to immediately notify the competent authority of the suspected presence of any of 10 diseases referred to in Annex I. The diseases listed in Annex I include rinderpest, bluetongue, and sheep and goat pox.

The last requirement under the heading of public, animal and plant health concerns the control and eradication of bluetongue. Annex II(B)(15) to the Regulation refers to Article 3 of Council Directive 2000/75/EC, which requires Member States to ensure the immediate, compulsory notification to the competent authority if circulation of the bluetongue virus is suspected or confirmed. The difference – if any – between Article 3 of Directive 2000/75/EC and Article 3 of Directive 92/119/EEC is that the latter requires notification of the 'of the suspected presence' of the disease, while the former requires notification 'if circulation of the bluetongue virus is suspected or confirmed'.

[5] Requirements for Animal Welfare

The cross compliance requirements relating to animal welfare are in Annex II(C)(16), (17) and (18) to Regulation (EC) No 73/2009, which refer to three different legislative

---

18. The wording of Annex II(B)(13) refers to Council Directive 85/511/EEC on measures for the control of foot-and-mouth disease. Directive 85/511/EEC has been repealed and replaced by the rules in Directive 2003/85/EC. According to Art. 91 in Directive 2003/85/EC, references to Directive 85/511/EEC in Union legislation are to be construed as references to Directive 2003/85/EC, and should be read in conjunction with the correlation table in Annex XX. The reference to Art. 3 in Directive 85/511/EEC should thus be read as a reference to Art. 3(1)(a) of Directive 2003/85/EC.
19. See §7.03 above on cross compliance requirements laid down in directives.

acts. Annex II(C)(16) refers to Articles 3 and 4 of Council Directive 2008/119/EC.[20] Article 3 lays down rules on the use of individual pens and the provision of minimum areas. Article 4 refers to Annex I to the Directive which contains detailed requirements for the rearing and feeding of calves.

Annex II(C)(17) refers to Articles 3 and 4 of Council Directive 2008/120/EC.[21] Article 3 lays down detailed rules on pens and installations for pig breeding and rearing. Article 4 supplements this with further requirements by a reference to the rules in Annex I.

Annex II(C)(18) refers to Article 4 of Council Directive 98/58/EC on the protection of animals kept for farming purposes. Article 4 requires the Members States to ensure that the conditions under which animals are bred or kept comply with the provisions set out in the Annex to the Directive. These provisions relate to staffing; supervision of the animals; record keeping of veterinary treatment and the number of mortalities; the animals' freedom of movement; the buildings and accommodation for the animals; special requirements for animals kept outside; the use of automated or mechanical equipment; requirements for feed, water and other substances; conditions for operations on animals; and requirements for breeding procedures.

[6]   Requirements for Agricultural Land (Good Agricultural and Environmental Conditions)

The requirements for agricultural land and environmental conditions are divided into two parts. First, Article 6(1) and Annex III to Regulation (EC) No 73/2009 draw together a number of standards for agricultural land and its maintenance in good agricultural and environmental condition. Second, Article 6(2) lays down some obligations for areas under permanent pasture.[22]

[a]   Requirements as to the Condition of the Land

Article 6(1) of Regulation (EC) No 73/2009 requires Member States to ensure that all agricultural land, especially land which is no longer used for production purposes, is

---

20. Annex II(C)(16) refers to Directive 91/629/EEC, which was repealed and replaced by the rules in Directive 2008/119/EC. According to Art. 12 of Directive 2008/119/EC, references to Directive 91/629/EEC in Union legislative acts are to be construed as reference to Directive 2008/119/EC and should be read in conjunction with the correlation table in Annex III to this Directive. The reference to Arts 3 and 4 of Directive 91/629/EEC should thus be read as a reference to Arts 3 and 4 of Directive 2008/119/EC.

21. Annex II(C)(17) refers to Directive 91/630/EEC, which was repealed and replaced by the rules in Directive 2008/120/EC. According to Art. 13 of Directive 2008/120/EC, references to Directive 91/630/EEC in Union legislative acts are to be construed as reference to Directive 2008/120/EC and should be read in conjunction with the correlation table in Annex III to this Directive. The reference to Arts 3 and 4 of Directive 91/630/EEC should thus be read as a reference to Arts 3 and 4 of Directive 2008/120/EC.

22. On the meaning of 'good agricultural and environmental condition' in connection with the earlier cross compliance requirements in Regulation (EC) No 1782/2003, see Catharina Meyer-Bolte, *Agrarrechtliche Cross Compliance als Steuerungsinstrument im Europäischen Verwaltungsverbund* 111-112 (Nomos, 2007).

maintained in a good agricultural and environmental condition. For this purpose, Member States are required to define, at national or regional level, minimum requirements for good agricultural and environmental condition on the basis of the framework established in Annex III to the Regulation. In this respect, Member States have to take into account the specific characteristics of the areas concerned, including soil and climatic conditions, existing farming systems, land use, crop rotation, farming practices, and farm structures. It is expressly stated in Article 6(1) that Member States may not define minimum requirements that are not provided for in the framework in Annex III.[23] The Commission has stated that the list in Annex III is closed and limited, as the 'issues or standards' in Annex III must be defined at Union level in order to ensure minimum and equal conditions for farmers from different areas and Member States.[24]

The further details in Annex III, about what requirements apply to agricultural land for it to be regarded as being in good agricultural and environmental condition pursuant to Article 6, are not given with reference to specific provisions in the legislation.[25] What Annex III does is to set out in tabular form five 'issues', and linked to each of these are broadly expressed 'compulsory standards' and 'optional standards'. For example, one issue is stated to be 'Soil organic matter', for which the compulsory standard is stated to be 'Arable stubble management' and the optional standard is stated to be 'Standards for crop rotations'. According to Article 6(1), the optional standards (referred to in Article 6(1) simply as 'standards') are voluntary unless the Member State has established such a standard as a minimum requirement prior to 1 January 2009 or the Member State uses national provisions for the standard.

The reference in Article 6(1) to the much broader standards in Annex III gives the Member States greater room for manoeuvre than the reference to Annex II in Article 5, which contains a concrete reference to the requirements and conditions of Union legislation.

[b] Requirements for Permanent Pasture

The second part of the requirements for agricultural land and environmental conditions are regulated in Article 6(2) of Regulation (EC) No 73/2009 and relate to areas of permanent pasture. The rule requires the Member States to ensure that land which was

---

23. In July 2007, the Commission wrote to the Member States that, under cross compliance, farmers cannot be sanctioned for disregarding obligations relating to issues or standards that are not included in Annex IV to Regulation (EC) No 1782/2003, then in force, corresponding to Annex III to Regulation (EC) No 73/2009; see European Court of Auditors, *Is Cross Compliance an Effective Policy?* 19 Special Report No 8/2008. On an obligation to maintain visible public rights of way and the requirement of 'good agricultural and environmental condition', see Case C-428/07 *Mark Horvath* [2009] ECR I-6355.
24. See the Commission's statement in the European Court of Auditors, *Is Cross Compliance an Effective Policy?* 45 Special Report No 8/2008.
25. The only exception to this is the establishment of buffer strips along water courses in connection with the protection and management of water. This requirement refers to the conditions for adding fertilizer to land near water courses in Council Directive 91/676/EEC. See §7.03[A][1] on Directive 91/676/EEC.

under permanent pasture at a given date is maintained under permanent pasture.[26] The aim of this general rule is to ensure that there is not a massive conversion of pasture land to arable land, and the justification for this rule is that such pasture land has a positive environmental effect.[27] The wording of the provision, especially the use of the word 'maintained', indicates that areas that were permanent pasture land on the relevant date cannot be converted to arable land. However, this is not the Commission's interpretation of the rule. In Article 3 of the Implementation Regulation, on the maintenance of land under permanent pasture, Member States must ensure that the ratio of the land under permanent pasture in relation to the total agricultural area does not decrease to the detriment of land under permanent pasture. The wording of Article 6(2) of Regulation (EC) No 73/2009 appears to lock in areas that were under permanent pasture on a given date, but Article 3 of the Implementation Regulation appears to give the Member States some scope for allowing permanent pasture to be converted to arable land, provided that other areas are converted to permanent pasture. This possibility of converting areas has been criticized for not protecting the quality of the existing permanent pastureland. It has been pointed out that this exception allows the Member States to reduce areas of high natural value and compensate for this by increasing pasture areas with low environmental value.[28] The Commission has rejected this criticism, referring to the purpose of Article 6 as being to avoid massive conversions of permanent pasture. According to the Commission, the Union legislator has laid down a rule that is purely quantitative and not qualitative. At the same time, the Commission has referred to the fact that the quality of permanent pasture is protected by other cross compliance requirements, including the Habitats Directive on the categorization of Natura 2000 areas.[29]

If it is found that the ratio of permanent pasture falls in a Member State, Article 4(1) of the Implementation Regulation requires the Member State concerned to provide that farmers applying for aid under any of the direct payment schemes may not convert land under permanent pasture without prior authorization.[30]

There are two exceptions to the general rule in Article 6(2) of Regulation (EC) No 73/2009 that Member States must ensure maintenance of permanent pasture areas. First, according to Article 6(2), second paragraph, a Member State may, in duly

---

26. In principle, the relevant date is the date provided for the area aid applications for 2003, but the date for the new Member States other than Bulgaria and Romania is 1 May 2004. For Bulgaria and Romania the date in question is 1 Jan. 2007. There are significant problems in determining how large the area of an individual Member State was permanent pasture on the given date, and Member States could be tempted to put the reference area at an unrealistically low level; see the European Court of Auditors, *Is Cross Compliance an Effective Policy?* 20 and 46 Special Report No 8/2008.
27. See recital 7 in Regulation (EC) No 73/2009.
28. See the European Court of Auditors, *Is Cross Compliance an Effective Policy?* 21 Special Report No 8/2008.
29. See the Commission's statement in the European Court of Auditors, *Is Cross Compliance an Effective Policy?* 46 Special Report No 8/2008.
30. However, see immediately below on the 10% rule in Art. 3(2) of the Implementation Regulation. See also Art. 4(2) of the Implementation Regulation for the situation where a Member State is unable to fulfil the obligation to maintain an area of permanent pasture of a certain size.

justified circumstances, derogate from the first paragraph, provided it takes action to prevent any significant decrease in its total permanent pasture area. The Commission has given expression to this exception in Article 3(2) of the Implementation Regulation by generally allowing Member States to reduce their area of permanent pasture by up to 10% compared to the ratio for the relevant reference year.

As the second exception to the general rule in Article 6(2), first paragraph, according to Article 6(2), third paragraph, the Member States may allow land under permanent pasture to be given up for the purpose of afforestation. However, this is on condition that the afforestation is compatible with the environment and excludes plantations of Christmas trees and fast growing species cultivated in the short term.

[B]   Cross Compliance in Connection with Support Pursuant to the Rural Development Regulation

The Rural Development Regulation contain only few reference to cross compliance requirements, but its Articles 50a and 51 contain requirements for cross compliance in connection with rural development support.

Article 50a states that a beneficiary receiving payments under Article 36(a)(i) to (v) and Article 36(b)(i), (iv) and (v) must respect the statutory management requirements and the requirement for good agricultural and environmental conditions provided for in Articles 5 and 6 of, and in Annexes II and III to, Regulation (EC) No 73/2009. The payments referred to under Article 36 of the Rural Development Regulation concern support for sustainable use of agricultural and forestry land. The requirements referred to in Regulation (EC) No 73/2009 apply to the whole holding. According to Article 50a(1), second paragraph, the obligation to comply with the statutory management (cross compliance) requirements and the requirements for good agricultural and environmental conditions do not apply to non-agricultural activities on a holding or to non-agricultural areas for which no support is claimed in accordance with Article 36(b)(i), (iv) and (v) of the Regulation.

If a recipient of support does not comply with the cross compliance requirements, then according to Article 51(1) there will be 'non-compliance'. The Rural Development Regulation does not itself define 'non-compliance', but it must be assumed to be defined in the same way as in Regulation (EC) No 73/2009.[31] According to Article 51, non-compliance that is directly attributable to the acts or omissions of the person who has sought support in connection with the relevant Article 36 measures can result in payments to that applicant being reduced or excluded.

---

31.   On this term, see §7.04[B] below.

§7.04 ENFORCEMENT AND SANCTIONS FOR FAILURE TO FULFIL THE CROSS COMPLIANCE REQUIREMENTS

[A] Controls of the Farmer

According to Article 23(1), first paragraph, of Regulation (EC) No 73/2009, a farmer's failure to comply with cross compliance requirements is referred to as 'non-compliance'.[32] With a view to ensuring fulfilment of these requirements, Article 22 of the Regulation requires Member States to carry out on-the-spot checks to verify whether a farmer complies with the obligations in Articles 4, 5 and 6 of the Regulation.[33] The content and extent of the control of farmers' cross compliance is set out by the Commission in Articles 50–54 of the Implementation Regulation. According to Article 50, the national authorities must carry out controls by performing on-the-spot-checks. The control is normally carried out of each farmer selected with a view to making sample on-the-spot-checks by a single control visit. The control consists of verification of compliance with the requirements and standards that can be checked at the time of the visit. The purpose of the control visit is both to verify compliance with the requirements and standards that can be checked at the time of the visit, and to see whether there are issues that require further control.[34] Sample on-the-spot-checks must be carried out within the same calendar year in which the aid applications are submitted.[35]

The control must cover the whole of the area of the agricultural holding. Inspection in the field as part of an on-the-spot check may be limited to a sample of at least half of the agricultural parcels concerned by the requirement or standard for the holding.[36] This requires that such a sample guarantees a reliable and representative level of control in respect of the requirements and standards. According to Article 53(2), if the sample check reveals non-compliance, the sample of agricultural parcels actually inspected must be increased.

---

32. On this non-compliance, see §7.04[B] below
33. With regard to the enforcement and sanctioning of cross compliance requirements associated with rural development support, see Art. 51 of the Rural Development Regulation. For further on controls in connection with the earlier Regulation (EC) No 1782/2003 which established the cross compliance arrangements, see Catharina Meyer-Bolte, *Agrarrechtliche Cross Compliance als Steuerungsinstrument im Europäischen Verwaltungsverbund* 161-202 (Nomos, 2007).
34. See Art. 53(3) of the Implementation Regulation. In Art. 2(34) and (35) of the Implementation Regulation it is emphasized that there is a distinction between 'requirements' and 'standards' for the purposes of cross compliance. 'Requirement' means each individual statutory management requirement resulting from any of the Arts referred to in Annex II to Regulation (EC) No 73/2009 within a given act. 'Standards' means the standards as defined by the Member States in accordance with Art. 6 of Regulation (EC) No 73/2009 and Annex III thereto.
35. See Art. 53(6). Art. 54 provides that for each on-the-spot-check the control authority must draw up a control report, and it gives detailed rules for the content of the control report.
36. On the Member States' obligation to introduce supervisory measures and inspection procedures, see Case C-418/06 P Belgium *v.* Commission [2008] I-3047. On prevention of on-the-spot-checks see Case C-536/09 *Marija Omejc* [2011] ECR and Case C-188/11 *Peter Hehenberger* [2012] ECR.

[B] Non-compliance

As stated above, according to Article 23(1) of Regulation (EC) No 73/2009, a farmer's failure to comply with cross compliance requirements is referred to as 'non-compliance'.[37] The terminology underlines that the Regulation lays down some administrative law conditions for individual farmers to receive public support.[38] If a farmer does not fulfil these conditions then, in terms of general administrative law, there will not be authority to pay support to that farmer. On the question of the right to public support, from the perspective of administrative law it is not relevant whether the farmer has given incorrect information to the relevant authority negligently or deliberately. If, objectively, the farmer does not fulfil the properly authorized conditions for the payment of support, the farmer has no right to the support and the authority may not pay it.

[C] Reduction of or Exclusion from Support Payments

According to Article 23(1) of Regulation (EC) No 73/2009, if it is found that statutory management requirements or the requirements for good agricultural and environmental conditions are not complied with at any time during a calendar year, then the support paid must be reduced or excluded. In the event of reduction, there are differing reduction percentages depending on the actual situation. Where the reduction affects direct payments, the percentage will apply to the total amount of direct payments which the farmer in question has received or should receive on the basis of an application for support which they have submitted during the calendar year in which non-compliance is established.[39]

The rules on the reduction of or exclusion from support payments are administrative sanctions, not criminal sanctions.[40] Cases concerning a reduction of or exclusion from support payments are independent of whether criminal proceedings are brought against a farmer. Whether or not disregarding requirements for cross compliance gives rise to criminal proceedings depends on the criminal law of the individual Member State.

If non-compliance leads to the reduction of or exclusion from support payments, it is expressly stated that the non-compliance in question must be due to an act or

---

37. In Case C-188/11 *Peter Hehenberger* [2012] ECR, it was held that where the beneficiary of agri-environmental support has prevented an on-the-spot check from being carried out, making it impossible to ascertain whether the conditions for eligibility of the aid have been complied with throughout the commitment period, the applications for agri-environmental aid concerned had to be rejected.
38. For comparison, the French language version of the Regulation uses the term '*non-respect*' and the German language version used the term '*Nichteinhaltung*'. In these versions the terminology used is in accordance with the ordinary terminology of administrative law. However, the Danish language version uses the term '*misligholdelse*', which refers to contract law, where this term is used in respect of a breach of a private agreement between two parties.
39. See Art. 70(8)(a) of the Implementation Regulation. There is a special rule for reductions where the reduction relates to support pursuant to the rules of the Rural Development Regulation on vines and grapes.
40. However, because of their penal character, in certain circumstances it can be questioned whether the reductions could be characterized as fines.

omission directly attributable to the farmer who submitted the aid application in the calendar year concerned. The fact that Article 23 refers to both acts and omissions emphasizes that the farmer can also be liable if the breach of a cross compliance requirement is not due to some act, but where he has failed to do what is necessary to fulfil the requirement in question. It must be assumed that the farmer is fully liable for any non-compliance which is due to workers employed on the farm. If one of the farmer's joint contractors does something which the farmer could not anticipate, and if taken in isolation this action leads to the non-fulfilment of cross compliance requirements, it cannot in principle be attributable to the farmer.[41]

The statement in Article 23 that the non-compliance must be 'attributable' to the farmer covers all forms of negligence and all degrees of intention. This is made clear in Article 24 on the implementation of reductions of or exclusions from support payments due to non-compliance with cross compliance requirements. Article 24(2), first paragraph, gives the reduction percentage applicable in the event of negligence, and Article 24(3) gives the reduction percentage applicable in the event of intentional non-compliance. What should be considered negligent and what should be considered intentional gives rise to special questions of evidence. Neither Regulation (EC) No 73/2009 nor the Implementation Regulation contains any rules on the evaluation of evidence, and this is thus a matter for the national authorities.[42]

It may seem a strict rule that simple negligence can lead to a reduction of support payments. However, Article 24(1) of Regulation (EC) No 73/2009 states that, as a general principle that is applicable to both negligent and intentional non-compliance, account must be taken of the severity, extent, permanence and repetition of the non-compliance.

In Article 47(2) to (4) of the Implementation Regulation, the Commission has defined what is meant by 'extent', 'severity' and 'permanence'. According to Article 47(2), the 'extent' of non-compliance is determined taking account whether the non-compliance has a far-reaching impact or whether it is limited to the farm itself.[43] According to Article 47(3), an assessment of the 'severity' of non-compliance should take into account the importance of the consequences of the non-compliance, given the aims of the requirement or standard concerned. And according to Article 47(4), whether a non-compliance is of 'permanence' depends on the length of time for which the effect lasts or the potential for terminating those effects by reasonable means.

Regulation (EC) No 73/2009, Article 24(2), second paragraph, adds a *de minimis* rule to the general principle in Article 24(1). In duly justified cases, Member States may

---

41. In Case C-11/12 *Maatschap L.A. en D.A.B. Langestraat* [2012] ECR, it was established that a farmer who receives aid (even if he does not actually enjoy possession of the parcel for which he has applied for the direct payment) must bear the risk of the attribution of a case of non-compliance with the cross compliance rules on that parcel due either to the negligence or the intentional act of the person to whom or from whom the transfer of that parcel was made.

42. In Denmark, the authorities have developed the practice whereby in principle a non-compliance can only be regarded as intentional when prior written attention is drawn to the rules in question during a period of three years prior to the inspection visit or if the authorities can prove that the rules have been clearly drawn to the attention of the farmer in one way or another.

43. On the meaning of a farm or agricultural holding, see §7.01 above.

decide to make no reduction where a case of non-compliance is to be considered as minor. Article 24(2), second paragraph, refers to cases where the *severity, extent and permanence* of non-compliance mean that the non-compliance should be considered minor, but unlike Article 24(1) this provision does not refer to the significance of the *repetition* of the non-compliance.[44]

As yet another rule limiting the consequences of negligent non-compliance, Article 24(2) provides that in the event of negligence the percentage of reduction shall not exceed 5% and, in the event of repeated non-compliance it shall not exceed 15%.[45] The Commission has given more concrete expression to this framework in Article 71 of the Implementation Regulation. As a general rule, with many exceptions, it is provided that the reduction shall be 3% of the total amount of the support payment.

In Article 47 of the Implementation Regulation, the Commission has defined what is meant by 'repeated' non-compliance. There is 'repeated' non-compliance when non-compliance with the same requirement, standard or obligation is determined more than once within a consecutive period of three calendar years. Repetition requires that the farmer has been informed of a previous non-compliance and has had the possibility to take the necessary measures to terminate that previous non-compliance.[46]

A farmer who intentionally fails to comply with cross compliance requirements must suffer a significant fall in his support payments. According to Article 24(3) of Regulation (EC) No 73/2009, the percentage of reduction is, in principle, at least 20% and may go as far as total exclusion from one or several aid schemes and apply for one or more calendar years. However, in cases of intentional non-compliance the authorities are obliged to make a concrete evaluation of the severity of the non-compliance. As stated above, account must be taken of the severity, extent, permanence and repetition of the non-compliance. But the wording of Article 24(3), that the percentage of reduction is, 'in principle', at least 20% also opens up for the possibility that an authority can in fact determine a percentage that is less than 20%.[47] The fact that paragraph 3 states that the sanction 'may go as far as' total exclusion suggests that total exclusion must be reserved for particularly serious cases on intentional non-compliance. It must be assumed that total exclusion is not applicable to cases of negligent non-compliance.

As a general rule which also concerns failure to fulfil cross compliance requirements, Article 31 provides that national authorities that control cross compliance must

---

44. Author's emphasis.
45. Not all language versions of Art. 24(2) are equally clearly expressed.
46. Article 47(1) refers to the same requirement, standard or the obligation referred to in Art. 4 of the Implementation Regulation. However, Art. 4 concerns the maintenance of land under permanent pasture at individual level, and it must be assumed that the intention is to refer to Art. 2, which contains definitions of various terms used in the Regulation, including 'requirement' and 'standards' (but not obligations).
47. 'In principle' is an ambiguous term, as it could be understood as meaning that the reduction may never be less than 20%.

recognize that there can be force majeure or exceptional circumstances in some situations. The provision gives five examples of such exceptional circumstances. They are:

(a) the death of the farmer;
(b) long term professional incapacity of the farmer;
(c) a severe natural disaster gravely affecting the holding's agricultural land;
(d) the accidental destruction of livestock buildings on the holding;
(e) an epizootic affecting part or all of the farmer's livestock, such as foot-and-mouth disease or BSE.

The wording of Article 31 underlines that these five situations must be considered as examples, but they are in line with the case law of the Court of Justice on force majeure.[48]

The total amount of reductions and exclusions for one calendar year may not be more than the total support payments that a farmer would have been entitled to if he had not failed to comply with the cross compliance requirements.[49]

---

48. See Ch. 4, §4.06[C], on force majeure and loss of security given in connection with licenses etc.
49. See Art. 24(4) of Regulation (EC) No 73/2009.

CHAPTER 8

Competition and State Aid

§8.01 THE RULES ON COMPETITION AND STATE AID IN THE TREATY ON THE FUNCTIONING OF THE EUROPEAN UNION

Article 3 of the Treaty establishing the European Community (TEC) gave a list of the aims of the European Community. Article 3(1)(g) stated that the activities of the Community included a system ensuring that competition in the internal market was not distorted. This introductory emphasis on the importance of regulating competition is not found in the TFEU. The preamble to the TFEU refers to the importance of fair competition, and its Article 3(1) states that the Union has exclusive competence to establish the competition rules necessary for the functioning of the internal market. The most important Treaty rules on competition are in Articles 101, 102 and 107. These rules prohibit three main types of activity. First, they prohibit agreements between two or more undertakings which distort competition. Second, they prohibit a single undertaking distorting competition by the abuse of its dominant position on a market. And third, they prohibit distortion of competition by an individual Member State's aid to undertakings located in that State.

The reason why the first two activities are prohibited by Articles 101 and 102 is that it is not only a State and its public authorities that can prevent or restrict the free movement of goods between Member States. Free movement can also be restricted by market operators.

In the agricultural sector, distortion of competition by agreements can be illustrated by the *Suiker Unie* case. A number of undertakings in the sugar sector, located in Germany, Italy and the Netherlands, had entered into agreements that resulted in the national markets for sugar being kept separate, even after the implementation of the market organization for sugar. In the market organization, the Union legislator had laid down rules which, in principle, ensured that there would be a single large market for sugar within the Union. However, the agreements between the sugar companies meant that sales from one country (i.e., one production area) to another had to be made via the undertakings concerned. The Court of Justice ruled that these agreements distorted competition and that they were contrary to the prohibition of agreements that distort trade that is now to be found in Article 101 TFEU.[1]

There are several ways in which competition can be distorted by the abuse of a dominant position on a market. Among other things, a refusal to supply a customer which is not a competitor to the supplier can be an abuse of a dominant position if the aim is to prevent the customer trading in the products of the supplier's competitor. In the agricultural sector, this can be illustrated by the *United Brands* case.[2] United Brands was a banana producer with a dominant position in the market for bananas in the internal market, delivering its bananas to authorized distributors in a number of Member States. One of United Brands' distributors participated in an advertising campaign for bananas from one of United Brands' competitors, whereupon United Brands stopped supplying the distributor in question. The Court of Justice ruled that, by its refusal to supply, United Brands had abused its dominant position contrary to Article 102. The Court stated that:

> an undertaking in a dominant position for the purpose of marketing a product – which cashes in on the reputation of a brand name known to and valued by the consumers – cannot stop supplying a long standing customer who abides by regular commercial practice, if the orders placed by that customer are in no way out of the ordinary.

The Court argued, among other things, that the refusal to sell would limit markets which would be to the disadvantage of consumers and would amount to discrimination which might ultimately eliminate a trading party from the relevant market.[3]

As stated, Article 107 imposes the third prohibition, that a Member State may not give aid to undertakings located in that State. If a State gives financial support to one of its 'own' undertakings, these undertakings would have a competitive advantage over undertakings located in other Member States which do not provide such support. In the agricultural sector, this can be illustrated by the example of the *Philip Morris* case.[4] The case concerned aid which the Dutch Government wanted to provide for a major tobacco producer, so that it could increase its production capacity. However, the Court of Justice held that the aid would distort competition, and that the conditions for allowing aid had in any case not been fulfilled. The Court stated that when State financial aid strengthens the position of an undertaking compared with competing undertakings trading within the Union, it must be assumed that trade is affected by that aid. In this case, the Dutch Government proposed to grant aid to an undertaking which exported a high percentage of its production to other Member States. The Court also

---

1. Joined Cases 40 to 48, 50, 54 to 56, 111, 113 and 114-73 *Suiker Unie* [1975] ECR 1663.
2. Case 27/76 *United Brands* [1978] ECR 207. In its judgment, the Court did not expressly address the question of whether Art. 102 TFEU applies to cases concerning agricultural law, but assumed that it did. In its submission the Commission touched on this and argued, with reference to the *Suiker Unie* case (referred to in footnote 1), that Art. 102 was applicable in this case; see 248 of the case report on the Commission's submission.
3. *Ibid.*, paras 182-183.
4. Case 730/79 *Philip Morris Holland* [1980] ECR 2671. Even though tobacco is unquestionably an agricultural product within the meaning of Union law, the rules of agricultural law were not involved in the case. On agricultural products, see Ch. 2, §2.02.

emphasized that the aid in question was to help to enlarge the undertaking's production capacity and thus its capacity to deliver to customers in other Member States. The aid would have reduced the cost of improving the production facilities and thereby give the undertaking a competitive advantage over manufacturers who have carried out or intend to carry out a similar increase in production capacity at their own expense.[5]

§8.02 THE COMPETITION RULES AND THE AGRICULTURAL SECTOR

In addition to securing undistorted competition, subparagraph (e) of Article 3(1) TEC referred to the fact that another of the Community's aims was the introduction of a common policy in the sphere of agriculture. As it was clear from the start that it would be difficult to reconcile these two aims in practice, a special rule on the application of the competition rules to agriculture was included in Article 36 TEC. Today this rule is in Article 42 TFEU.

The difficulty of balancing regard for the agricultural sector and regard for competition came to a head in a case involving a claim for annulment brought by Germany against the Council in 1993. Germany sought the annulment of parts of a regulation which established the then market organization for bananas. One of Germany's claims was that the provisions of the regulation on the tariff quota for imports and the way quotas were allocated was contrary to Article 3(1)(g) TEC, with its prohibition of the distortion of the internal market.[6] This claim was dismissed by the Court of Justice which stated that the prevention of the distortion of competition was not the only aim referred to in Article 3, and the provision also referred to the common agricultural policy. The Court stated that: 'The authors of the Treaty were aware that the simultaneous pursuit of those two objectives might, at certain times and in certain circumstances, prove difficult', and for this reason what is now Article 42 TFEU was included in the Treaty. On this basis, with reference to Article 42 the Court stated that: 'Recognition is thus given to both the priority of the agricultural policy over the objectives of the Treaty in the field of competition and the power of the Council to decide to what extent the competition rules are to be applied in the agricultural sector'.[7]

Pursuant to Article 42 TFEU, if the Union legislator decides that the competition rules should apply to the production of and trade in agricultural products, the legislator must take into consideration the requirements of competition policy.[8]

As stated above in section §8.01, the TFEU competition rules are concerned with agreements between undertakings that distort competition, undertakings' abuse of their dominant position, and State aid to undertakings. According to Article 42(1) TFEU, these rules apply to production of and trade in agricultural products only to the extent determined by the European Parliament and the Council. This is in contrast to the rules on the free movement of goods, as according to Article 38(2) TFEU, these rules apply to agricultural products unless otherwise provided in Articles 39–44 TFEU.

---

5. *Ibid.*, para. 11.
6. See §8.01 above on Art. 3(1)(g) TEC, and Ch. 4, §4.06[A][3], on tariff quotas.
7. Case C-280/93 *Germany v. Council* [1994] ECR I-4973, paras 58–61.
8. Case 131/86 *United Kingdom v. Council* [1988] ECR 905.

Decisions of the European Parliament and the Council pursuant to Article 42(1) TFEU, that the competition rules are to apply, are made pursuant to the legislative procedure set out in Article 43(2) TFEU.[9] Under Article 42, second paragraph, on a proposal from the Commission, the Council can authorize the granting of specific forms of aid. This relates to aid for the protection of enterprises handicapped by structural or natural conditions, and aid within the framework of economic development programmes.[10]

The Union legislator has exercised the authority under Article 42(1) TFEU to adopt Council Regulation (EC) No 1184/2006 applying certain rules of competition to the production of, and trade in, agricultural products, and these rules still apply following the entry into force of Council Regulation (EC) No 1234/2007 establishing a common organization of agricultural markets and on specific provisions for certain agricultural products (the 'Single CMO Regulation').[11] Regulation (EC) No 1184/2006 largely carries forward the older EEC Council Regulation No 26/62. According to the very short *travaux préparatoires* to the current Regulation, its purpose is to codify EEC Council Regulation No 26/62, and it does not involve any substantive amendments to the competition law regulation of agriculture laid down in that Regulation.[12] Article 4 of Regulation (EC) No 1184/2006 expressly states that references to Regulation No 26/62 in other legislation are to be construed as references to Regulation (EC) No 1184/2006.

Before the Commission put forward its proposal for what later became Regulation No 26/62, it had drawn up an overall proposal for a common agricultural policy. In this connection it had emphasized the necessity of the competition rules applying to the production of and trade in agricultural products.[13] However, when later in the same year the Commission made its proposal for the Regulation, it stated that the competition rules, in the form of the prohibition of cartels, abuse of dominant position and the prohibition of dumping, could only apply to the extent that they did not restrict national market organizations for agricultural products and did not put at risk the achievement of the goals of the common agricultural policy set out in Article 39 of the EEC Treaty.[14]

---

9.   On the legislative procedures for agricultural law, see Ch. 3, §3.05.
10. Prior to the Lisbon Treaty the Council alone adopted legislation on the application of the competition rules in the agricultural sector; see Art. 36 TEC.
11. Regulation (EC) No 1184/2006 entered into force on 21 Aug. 2006; see its Art. 5. Article 200 of Council Regulation (EC) No 1234/2007 on the common market organization changed a few provisions in Regulation (EC) No 1184/2006, has been amended several times since its entry into force, most recently by Council Regulation (EC) No 491/2009 in connection with the inclusion of the wine sector in the common market organization.
12. See the Commission's Proposal for a Council Regulation applying certain rules of competition to production of and trade in agricultural products (COM(2005) 613 final), 5; The Report of the European Parliament's Committee on Legal Affairs, A6-0121/2006 final, 5; and the Resolution adopted by the European Parliament, P6_TA(2006)0146. See also recital 1 of Regulation (EC) No 1184/2006.
13. See the Commission's proposal of 30 Jun. 1960, *Vorschlage zur Gestaltung und Durchfuehrung der gemeinsamen Agrarpolitik gemass Artikel 43 des Vertrages zur Gruendung der europäischen Wirtschaftsgemeinschaft*, document VI/COM(60)105, second part, 20, where it was stated, among other things, that a common agricultural policy would involve the introduction of 'gemeinsame Wettbewerbsregeln'.
14. See the Commission's proposal of 28 Nov. 1960, *Proposition d'un premier reglement du conseil concernant l'application de certaines règles de concurrence à la production et au commerce des*

Thus, the situation today is that Article 42 TFEU states that the competition rules apply to the agricultural sector only to the extent laid down in Union legislation. Regulation (EC) No 1184/2006 lays down general rules on the application of the competition rules of the TFEU to the agricultural sector, but at the same time special rules can be laid down under the Single CMO Regulation or other Union legislation.

Even though Regulation (EC) No 1184/2006 and the Single CMO Regulation make the Treaty competition rules applicable to the agricultural sector in principle, they also restrict the application of the competition rules and undertakings can thus have an interest in claiming to be covered by the rules of one of the two regulations. However, it is a condition for this that the individual undertaking is engaged in the production of or trade in what the TFEU considers to be agricultural products.[15] In the *Stremsel-en Kleurselfabrik* case, a cooperative company sought unsuccessfully to avoid the application of the Treaty competition rules under Council Regulation No 26/62 (corresponding to Regulation (EC) No 1184/2006) by claiming that animal rennet is an agricultural product.[16]

Article 1 of Regulation (EC) No 1184/2006 states that the Regulation lays down rules for the application of the competition rules in connection with the production of or trade in the products listed in Annex I to the TFEU, with the exception of the products covered by the Single CMO Regulation.[17] Article 175 of the Single CMO Regulation lays down the presumptive rule that Articles 101–106 TFEU and their implementing provisions apply to all agreements, decisions and practices referred to in Articles 101(1) and 102 TFEU which relate to the production of or trade in the products covered by the Single CMO Regulation. According to Article 175, this presumptive rule applies unless otherwise provided in the Single CMO Regulation. This legislation means that the rules in Regulation (EC) No 1184/2006 on the application of the competition rules to the agricultural sector only apply to the production of and trade in agricultural products that are not covered by the Single CMO Regulation. Within the area to which the Single CMO Regulation applies, which means much the most important agricultural sectors, it is Articles 175–182a of the Single CMO Regulation which determine the extent to which the competition rules in Articles 101–106 TFEU apply.

In general, a systematic distinction is made between the competition rules for undertakings and the State aid rules, and this distinction is also applied in the Single

---

produits agricoles en vertu de l'article 42 du traité, third recital, document VI/COM(60)160 final, where there is a reference to '*les règles de concurrence relatives aux ententes, à l'exploitation abusive des positions dominantes et aux pratiques de dumping*'.

15. On the meaning of 'agricultural products', see Ch. 2, §2.02.
16. Case 61/80 *Coöperatieve Stremsel-en Kleurselfabrik* [1981] ECR 851, paras 19–21. For cases where undertakings have unsuccessfully claimed that they are covered by the agricultural rules in order to avoid the application of the competition rules, see Case 123/83 *BNIC* [1985] ECR 391, paras 14–15 (cognac is not an agricultural product); Case T-61/89 *Dansk Pelsdyravler-forening* [1992] ECR II-1931, paras 36–39 (skins and pelts are not agricultural products); and Case C-250/92 *DLG* [1994] ECR I-5461, paras 21–24 (fertilizer and plant protection substances are not agricultural products).
17. On the scope of the Single CMO Regulation, see Ch. 4, §4.02[B].

CMO Regulation, Articles 175–179 of which lay down rules for undertakings, and Articles 180–182a lay down rules for State aid to undertakings.[18]

§8.03   COMPETITION RULES FOR UNDERTAKINGS IN THE AGRICULTURAL SECTOR: ARTICLES 101 AND 102 TFEU

[A]   Council Regulation (EC) No 1184/2006 Applying Certain Rules of Competition to the Production of, and Trade in, Agricultural Products

[1]   The Main Rule

Article 1a provides that, within the area to which the Regulation applies, Articles 101–106 TFEU and provisions for their implementation apply to all agreements, decisions and practices referred to in Articles 101(1) and 102 TFEU relating to the production of or trade in agricultural products.[19] However, this provision is laid down with the significant addition that it is subject to the provisions in Article 2.

Article 2(1) contains several exceptions to the main rule in Article 1a, but it only refers to Article 101(1), and not Article 102. This means that Article 102 applies without limit within the area covered by Regulation (EC) No 1184/2006.[20]

In its proposal for EEC Council Regulation No 26/62, the Commission proposed that Article 2 should state that neither the Treaty's prohibition of agreements that restrict competition nor the prohibition of the abuse of a dominant position should apply to agreements etc. that are part of a national market organization or which are necessary for the achievement of the goals of the common agricultural policy.

The European Parliament found that the prohibition in the Treaty of abuse of a dominant position should in principle apply to the agricultural sector, and thus proposed that this reference to the Treaty provision should be removed from Article 2.[21]

---

18. The competition rules of the EEA do not apply to agricultural products. According to its Art. 8(3), the EEA Agreement only applies to the product groups in the provision, and agricultural products are not listed.
19. For a definition of 'agricultural products' in connection with Regulation (EC) No 1184/2006, see §8.02 above.
20. See Usher, *EC Agricultural Law* 14 (Oxford University Press, 2d ed., 2001), where it is stated with reference to Art. 2, that: 'this effectively means that arts [101] to [106] of the Treaty are read subject to the specific provisions of the legislation establishing common organizations of agricultural markets'; however, see also p. 16. Cf. McMahon, *EU Agricultural Law* 16 (Oxford University Press, 2007): 'The exception in Article 2 relates only to Article [101]; Article [102] applies with full force to the production of or trade in agricultural products'. For a view that Art. 102 applies to an unlimited extent, see also Bellamy & Child, *European Community Law of Competition* 1176 (Oxford University Press, 6th ed., Roth & Rose 2008), who argue that Art. 2(1) must be interpreted restrictively as an exception to 'the general rule on competition'.
21. See the proposed amendment of 20 Jan. 1961 from Parliament's Agricultural Committee, for which further reasons were not given; Sitzungsdokumente 1960–1961, document 111, APE 5086, 5.

When the Council dealt with the Commission's proposal, in the first instance there was disagreement about whether the European Parliament's proposal should be followed. Belgium and Italy supported the Commission's proposal for a broad exception to the main rule in the Regulation, while Germany was the only Member State that expressly supported the European Parliament's proposal. Agreement was later reached, with the assent of the Commission, to follow the European Parliament's proposal.[22] As a consequence Council Regulation No 26/62 took a form which meant that the Treaty prohibition of the abuse of dominant position has unlimited effect in the agricultural sector, and continues to do so under Regulation (EC) No 1184/2006.[23] In the *United Brands* case, which concerned an undertaking's abuse of its dominant position, in accordance with the wording of Article 2 and with the *travaux préparatoires*, the fact that the case concerned an agricultural product (bananas) had no effect on the application of Article 102 TFEU.[24]

The Court of Justice has stated that, as an exception, Article 2 of Council Regulation No 26/62 must be interpreted restrictively, and the same must also apply to the rule in Regulation (EC) No 1184/2006.[25]

[2] The First Exception: An Integral Part of a National Market Organization

According to Article 2(1) of Regulation (EC) No 1184/2006, the first exception to the main rule is that Article 101(1) TFEU does not apply to the agreements, decisions and practices referred to in Article 1a which constitute an integral part of a national market organization.[26] In Chapter 4, section §4.01, it is pointed out that today it is only possible to establish or maintain national market organizations in very few areas, either because the area is not subject to regulation under Union law, or because Union law allows national market organizations. The scope of and practical consequences of the exception for national market organizations in Article 2(1) are thus very limited.

[3] The Second Exception: Necessary for Attainment of the Objectives Set out in Article 39 TFEU

As the second exception to the main rule in Article 1a, Article 3(1) provides that Article 101(1) TFEU does not apply to such agreements, decisions and practices as are

---

22. In the first instance, discussions took place in a working party of the Special Committee for Agriculture. For the initial disagreement of the working party, see the Report of 25 Feb. 1961, document S/84/61 (CSA 6), 3. For the subsequent disagreement, see the minutes of the working party of 26 Sep. 1961, document /431/1/61 (CSA 31 rev.), Annex p. 1.
23. See Joseph A. McMahon, *EU Agricultural Law* 16 (Oxford University Press, 2007); and Whish, *Competition Law* 964 (Oxford University Press, 7th ed., 2011). See also Bellamy & Child, *European Community Law of Competition* 1176 (Oxford University Press, 6th ed., Roth & Rose 2008).
24. Case 27/76 *United Brands* [1978] ECR 207.
25. See e.g., Case C-265/97 P *Florimex* [2000] ECR I-2061, para. 94.
26. See Ch. 4, §4.01, for a definition of a national market organization.

necessary for attainment of the objectives set out in Article 39 TFEU.[27] By the words 'necessary for attainment of the objectives set out in Article [39] of the Treaty', Article 2 of Regulation (EC) No 1184/2006 means that Article 101(1) TFEU must be read with a reservation in respect of the comprehensive regulation of agriculture that establishes the market organization. The earlier market organizations largely had the aim of removing or limiting price competition. In line with earlier market organizations, the Single CMO Regulation must be assumed expressly or tacitly to serve the purposes of Article 39, so that the parties to an agreement cannot claim that the agreement which limits competition is necessary for the attainment of the objectives set out in Article 39 of the Treaty.[28] According to the established case law of the Court of Justice, the establishment of a common market organization binds the Member States not to make exceptions to the organization or otherwise set it aside.[29] The same applies to undertakings.

There has been a question as whether, according to Article 2(1) of the Regulation, it is sufficient if merely one or some of the objectives of Article 39 TFEU are fulfilled. In the *Frubo* case, the Court of Justice established that all the objectives set out in Article 39 must be served by the agreement, decision or practice in question in order for the exception in Article 2(1) to be applicable. In a decision the Commission had ruled that an agreement between the members of an association of fruit importers and wholesalers (Frubo) restricted competition and thus infringed Article 101(1) TFEU. The association brought the case before the Court of Justice, claiming that the Commission's decision should be annulled, arguing among other things that the exception in Article 2 of Council Regulation No 26/62, then in force, was applicable. That provision, like the current rule in Article 2 of Regulation (EC) No 1184/2006, referred to the objectives set out in Article 33 [now 39] of the Treaty. Frubo argued that it was sufficient if some of the objectives in Article 39 were served. However, the Court of Justice rejected this interpretation of Article 2, stating that: 'the applicants have not shown in what respect their agreement, which is concerned with products coming from third countries, can be necessary to "increase agricultural productivity" or to "ensure a fair standard of living for the agricultural community"'.[30]

---

27. See Ch. 2, §2.05, on the interpretation of the aims stated in Art. 39 TFEU.
28. See Whish, *Competition Law* 966 (Oxford University Press, 7th ed., 2011): 'In practice it is likely that the Commission and the EU Courts will hold that the objectives of Article 39 are expressly or impliedly advanced by the provisions of any particular regulation establishing a common organisation of an agricultural sector, with the result that there is no remaining latitude for the parties to an agreement to argue that their agreement will have this effect.' For the same view, see Van Bael & Bellis, *Competition Law of the European Community* 1458 (Kluwer, 4th ed., 2005).
29. See e.g., Case 212/87 *Unilec* [1988] ECR 5075, para. 15; and Case 218/85 *Le Campion* [1986] ECR 3513, para. 12.
30. Case 71/74 *Frubo* [1975] ECR 563, paras 22–27. The case concerned as mentioned above Art. 2(1) of Council Regulation No 26/62, then in force; this rule has been carried forward by Regulation (EC) No 1184/2006. See also, on the emphasis of the requirement for the fulfilment of all objectives in Art. 39 TFEU, Case C-399/93 *Oude Luttikhuis* [1995] ECR I-4515, paras 25. Compare with the case law of the Court of Justice on the lawfulness of Union legislation on agricultural questions where the Court has allowed that the fulfilment of some of the objectives in Art. 39 TFEU is sufficient; see Ch. 2, §2.05[A].

The requirement, that all the objectives set out in Article 39 TFEU must be served, must also be taken seriously by the Commission. This was underlined in the *Florimex* case. The Commission had adopted a decision that the rules laid down or agreements entered into by a Dutch association of growers of flowers and ornamental plants were covered by Article 101(1) TFEU. However, in the same decision the Commission had found that these rules and agreements were necessary for the attainment of the objectives set out in Article 101(1) TFEU, and that the exception in Article 2 of Council Regulation No 26/62 applied. In connection with proceedings for annulment, in which the Commission's decision was challenged, the Court of Justice ruled that the Commission had not complied with its obligation by showing 'how the agreement between the members of a cooperative satisfies each of the objectives of Article 39 ... which must be interpreted strictly'.[31]

Prior to the entry into force of the Single CMO Regulation, the individual market organization, i.e., the regulation which established the organization in question, generally stated the means or instruments that should be used with a view to attaining the objectives in Article 39 TFEU.[32] Following the entry into force of the Single CMO Regulation, and its repeal of the various market organizations, these instruments are now found in the Single CMO Regulation. This means that the second exception in Article 32(1) can only be relied upon where the application of Article 101(1) TFEU will restrict or prevent one of the instruments referred to in the Single CMO Regulation and thus hinder the attainment of the objectives in Article 39. Where the Single CMO Regulation applies, it is difficult in practice to demonstrate that an agreement is necessary for the fulfilment of the objectives in Article 39, unless the Single CMO Regulation expressly or tacitly provides for such a kind of agreement.[33]

The *Cauliflower* case, dating from prior to Regulation (EC) No 1184/2006 and prior to the Single CMO Regulation, illustrates the connection between the exception in Article 2(1) and the means referred to in the relevant market organization. Some French organizations of producers of cauliflowers, artichokes and potatoes, on the one hand, and of wholesale and distribution undertakings on the other hand, had entered into an agreement which restricted the possibilities of the latter undertakings buying goods from others than the named producer organizations. The market organization for fruit and vegetables, which included cauliflowers, among other things, laid down rules for producers for the attainment of the objectives in Article 39, but not for these producers' trading partners, including the wholesale and distribution undertakings in question. Since the market organization did not lay down rules or instruments whereby

---

31. Case C-265/97 P *Florimex* [2000] ECR I-2061, para. 94.
32. See e.g., Commission Decision 86/596/EEC on agreements that distorted competition in the Dutch milk sector, point 54 (OJ L 348, 10 Dec. 1986, 50), where the Commission referred to the means that may be employed to attain the objectives in Art. 39 TFEU according to the market organization for milk and milk products then in force; see Regulation (EEC) No 804/68.
33. On the legal situation prior to the Single CMO Regulation, and while Council Regulation No 26/62 (the predecessor to Regulation (EC) No 1184/2006) applied, see Bellamy & Child, *European Community Law of Competition* 1204–1205 (Oxford University Press, 6th ed., Roth & Rose 2008); and Ratliff, *Agricultural Law for the European Union* 37 (43–49) (Academy of European Law and Irish Centre for European Law, Heusel & Collins eds.,1999).

the Article 39 objectives should be attained by wholesale and distribution undertakings, the exception in Article 2(1) was not applicable, but the main rule in Article 1 was applicable, whereby the agreement in question was subject to Article 101(1) TFEU.[34]

[4]   The Third Exception: Agreements etc. between Farmers, Farmers' Associations, or Associations of Farmers' Associations

Article 2(1), second sentence, of Regulation (EC) No 1184/2006 emphasizes that Article 101(1) TFEU in particular does not apply to agreements, decisions and practices of farmers, farmers' associations, or associations of such associations belonging to a single Member State, where there is no obligation to charge identical prices and which concern the production or sale of agricultural products or the use of joint facilities for the storage, treatment or processing of agricultural products. This is provided the Commission does not find that competition is thereby excluded or that the attainment of the objectives of Article 39 TFEU is jeopardized. This provision is a good example of convoluted legislation.[35]

For a time it was unclear, and it is still a matter for debate, whether this part of Article 2(1) is merely giving an example of the second exception described in section §8.03[A][3] above, or whether it is a third, independent exception.[36] In particular, the discussion has centred on the wording of the second sentence (see 'In particular, it shall not apply'), and the *travaux préparatoires*. Some have argued that the wording shows that these are merely special exceptions to the first paragraph, while others have attached more weight to the *travaux préparatoires*.[37]

---

34. See Commission Decision 78/66/EEC of 2 Dec. 1977 relating to a proceeding under Art. 85 of the EEC Treaty (IV/28.948 – Cauliflowers), Part III, point 2.
35. In Bellamy & Child, *European Community Law of Competition* 1178 (Oxford University Press, 6th ed., Roth & Rose 2008), this rule is characterized as being 'ambiguously drafted'.
36. Whish, *Competition Law* 964, n. 11 (Oxford University Press, 7th ed., 2011), rejects the idea that Art. 2(1) contains a third exception: 'Article 2 contains a sentence dealing with the activities of farmers' associations; this is not a further exception, but rather an embellishment of the policy expressed in that provision'. In support of his view, he refers to Commission Decision 85/76/EEC of 7 Dec. 1984 relating to a proceeding under Art. 85 of the EEC Treaty (IV/28.930 – Milchförderungsfonds), points 21–22; and Commission Decision 88/491/EEC of 26 Jul. 1988 relating to a proceeding pursuant to Art. 85 of the EEC Treaty (IV/31.379 – Bloemenveilingen Aalsmeer), points 150–152. On the Commission's decisions, see §8.03[A][4][b] below. For the same view as Whish, see de Cockborne, 'Les Règles communautaire de concurrence applicables aux entreprises dans le domaine agricole', RTDE 293 (306) (1988); and the Opinion of Advocate General Tesauro in Joined Cases C-319/93, C-40/94 and C-224/94 *Dijkstra* [1995] ECR I-4471, point 18. See §8.03[A][4][c] below on the *Dijkstra* case.
37. For arguments based on the significance of the *travaux préparatoires*, see de Cockborne, 'Les Règles communautaire de concurrence applicables aux entreprises dans le domaine agricole', RTDE 293 (305–306) (1988): '*La deuxième phrase de l'article 2, paragraphe 1, du règlement n° 26, qui ne figurait pas dans la proposition originale de règlement de la Commission, a été introduite à la demande du Parlement européen, pour protéger les coopérative agricoles*'; and Deringer, *The Competition Law of the European Economic Community* 372–373 (Commerce Clearing House, 1968).

[a]   The Travaux Préparatoires to Article 2(1), Second Sentence

It is stated in section §8.02 above that Regulation (EC) No 1184/2006 to a large extent carries forward Council Regulation No 26/62. Article 2(1) repeats the rule in the former Regulation word for word.

The Commission's proposal for the earlier Regulation did not contain the second sentence of the first paragraph, merely the two exceptions referred to in sections §8.03[A][2] and §8.03[A][3] above. When the draft was sent to the European Parliament for consultation pursuant to Article 37(2), third section, EC, the Parliament stated that the draft must be understood in the sense that the Treaty competition rules do not restrict agricultural cooperatives. The European Parliament did not propose that the wording of the Regulation should state that cooperative companies should have a special position in competition law, but in its response to consultation it restricted itself to stating: *'mais celle-ci aimerait toutefois s'assurer que l'application des articles 85 à 90 du traité ne portera pas préjudice à l'activité des coopératives agricoles'.*[38] The second sentence was drafted in negotiations in the Council in connection with its final adoption of the Regulation. France and Germany supported an express exception for farmers' associations, and Germany put forward a specific proposal for how the exception could be worded.[39]

The German proposal was supported by all the members of the Council except the Netherlands. The Commission also did not support the proposal.[40] The Netherlands argued that the proposal was superfluous, as Article 2(1) of the Commission's proposal not only referred to national market organizations, but also excluded agreements etc. that are necessary for the attainment of the Treaty objectives for the common agricultural policy. In the view of the Netherlands, the reference to the latter category of agreements was so broadly worded that regard for cooperative companies could not justify changing the Commission's proposal. The Netherlands stated that there was no doubt that cooperative companies were covered by the term 'market organization', and that the German proposal caused serious problems for the Dutch Government. The

---

38. See the Commission's proposal, with the Parliament's proposed amendments, without the exception in the second sentence of the first paragraph, Amtsblatt 1961, 361. See the emphasis on the importance of farmers' associations in the report of the European Parliament Committee on Agriculture, referred to in Parliament's proposals for amendments; Report of 11 January 1961, *Rapport intérimaire*, of the European Parliament Committee on Agriculture, document 107, in connection with Art. 1 of the draft regulation. Arts 85–90 of the EEC Treaty correspond to Arts 101–106 TFEU. Several Members of the European Parliament subsequently sought an answer from the Commission on whether it would protect the interests of farmers' associations in connection with the final adoption of Council Regulation No 26/62; see the written question to the Commission No 62, Amtsblatt 1961, 1637; and written question to the Commission No 74, Amtsblatt 1962, 161.
39. For the French position, see the working party of the Special Committee on Agriculture, minutes of 13 Sep. 1961, document S/431/61 (CSA 31), Annex p. 1, where information is given about France's position on the draft for Art. 2(1): *'wünscht jedoch eine Verbessrung des Vorschlags dahingehend, dass die landwirtschaftlichen Genossenschaften ausdrücklich in die Ausnahmeregelung einbezogen werden'.* For the German proposal, see the Special Committee on Agriculture, 6 Oct. 1961, document S/484/61 (CSA 36) Ausz. 1, p. 1.
40. *Extrait du procès verbal de la 59ème session du conseil de la C.E.E., tenue a Bruxelles, le 12 decembre 1961*, document 1690/61, p. 12, pp. 12-13.

distinction between associations, cooperatives and cartels was completely unknown in Dutch law.[41]

The Commission opposed the German proposal on four grounds. First, it referred to the fact that the Treaty provision, which prohibited agreements that restricted competition, only concerned trade between Member States, and that agricultural cooperatives only seldom traded between the Member States. According to the Commission, the great majority of cooperatives in the agricultural sector were not covered by the Treaty competition rules.[42] Second, the Commission said that it would be difficult to distinguish between cooperatives which would be exempt from the competition rules under the German proposal and other associations that would not be exempt.[43] Third, the Commission argued that the Treaty provision that prohibited agreements that restrict competition authorizes the Commission to grant specific exceptions to the prohibition.[44] Finally, as did the Netherlands, the Commission pointed out that the Commission's proposal for Article 2(1) made an exception for cooperative companies which enter into agreements etc., which are part of a national market organization or which are associated with the objectives of the agricultural policy set out in the Treaty.[45]

The Dutch opposition to the German proposal was maintained to the end, but finally the Dutch Minister agreed to the German proposal as a concession (*à titre de concession*).[46] When the Netherlands changed its position, the Commission realized that the Council could adopt the German proposal unanimously, and thereafter abandoned its opposition to the proposal.[47]

---

41. See the Special Committee on Agriculture, 6 Oct. 1961, document S/484/61 (CSA 36) Ausz. 1, p. 1; and *Extrait du proces verbal de la 59ème session du conseil de la C.E.E., tenue a Bruxelles, le 12 decembre 1961*, document 1690/61, p. 12.
42. The Commission referred to this fact both in the Special Committee and in the Council; see the Special Committee on Agriculture, 6 Oct. 1961, document S/484/61 (CSA 36) Ausz. 1, p. 1; and *Extrait du proces verbal de la 59ème session du conseil de la C.E.E., tenue a Bruxelles, le 12 decembre 1961*, document 1690/61, p. 12.
43. *Extrait du proces verbal de la 59ème session du conseil de la C.E.E., tenue a Bruxelles, le 12 decembre 1961*, document 1690/61, p. 13.
44. See the Special Committee on Agriculture, 6 Oct. 1961, document S/484/61 (CSA 36) Ausz. 1, p. 3. In the negotiations in the Council, at one point the Commission proposed as a compromise that a rule might be considered whereby agreements between cooperatives might be allowed up to the point where the Commission might determine that there was a distortion of competition; see *Extrait du proces verbal de la 59ème session du conseil de la C.E.E., tenue a Bruxelles, le 12 decembre 1961*, document 1690/61, p. 13. The Netherlands later found that the Commission had abandoned going further with this proposal; see *Extrait du proces verbal de la reunion restreinte tenue a l'occassion de la 60ème session du conseil de la C.E.E., a Bruxelles, du 18 au 22, les 29 et 30 decembre 1961 et du 4 au 14 janvier 1962, document R/78/62 (2ème partie)*, p. 312.
45. *Extrait du proces verbal de la 59ème session du conseil de la C.E.E., tenue a Bruxelles, le 12 decembre 1961*, document 1690/61, p. 13.
46. See *Extrait du proces verbal de la reunion restreinte tenue a l'occassion de la 60ème session du conseil de la C.E.E., a Bruxelles, du 18 au 22, les 29 et 30 decembre 1961 et du 4 au 14 janvier 1962, document R/78/62 (2ème partie)*, p. 312.
47. *Ibid.* 312. For the Commission's proposal of 30 Dec. 1961, containing the German proposal, see document VI/KOM(61)212 endg. 2. In the Council's revised draft of the Regulation of 8 February, document 18/1/62 (AGRI 5 rev.), a new (fourth) recital was included: 'Whereas special attention is warranted in the case of farmers' organisations'.

The German proposal had been subject to certain adjustments on the way, but these changes did not alter the fact that the purpose of including the second sentence in Article 2(1) of the Regulation was to create an independent exemption for agricultural cooperatives from the prohibition of agreements that restrict competition.[48]

[b] The Commission's Interpretation

It must be admitted that the wording of Article 2(1) is not clear, and while a review of the provision's *travaux préparatoires*, as above, is a substantial aid to the interpretation of the provision, there have been significant questions associated with its interpretation.[49] This doubt has been fuelled, not least, by the view which the Commission adopted later, after the adoption of Council Regulation No 26/62. Even though the Commission had accepted the proposal to amend the draft in connection with the adoption of the Regulation, ensuring the exemption of agricultural associations from the Treaty prohibition of agreements that restrict competition, later in 1984 in the *Milchförderungsfonds* case it expressly rejected the idea that Article 2(1), second sentence, gave agricultural cooperatives a special competitive position. The case concerned the setting up of a German fund which, as its title suggests, had the purpose of promoting the sales of milk. The fund was set up by German farmers' organizations and had, as its sole beneficiary, a partnership set up by the farmers' organizations together with their regional branches. The fund's assets were derived from payments collected by cooperative and private dairies on the basis of a given amount per litre of milk supplied. In some cases the fund paid out subsidies in connection with exports of German milk to other Member States. As part of its assessment of the activities of the fund, the Commission stated that Article 2(1), second sentence, was a special circumstance in relation to the first paragraph, and 'requires that the case satisfy one of the two conditions for the exception in the first sentence'. However, the Commission cautiously added that: 'even if the special circumstances described in the second sentence were intended to constitute a separate exception, this additional independent test would still not be met'.[50]

---

48. As stated above, the German proposal was presented at a meeting of the Special Committee on Agriculture, 6 Oct. 1961. At a later meeting of the Special Committee, Germany proposed making an adjustment to its original proposal; see Special Committee on Agriculture, 13 Oct. 1961, document S/431/4/61 (CSA 31 rev. 4), Annex p. 1. At a meeting of the working party it was decided to make a further change, against the protest of the Commission, see the minutes of the working party of 13 Dec. 1961, document S/644/61 (CSA 68), 2. With these additions, and with some linguistic polishing, the German proposal was included as the second sentence in Art. 2(1). In Groeben, von Boeckh, Thiesing & Ehlermann, *Kommentar zum EWG-Vertrag* 1372 (Nomos, 3rd ed., 1983), Schröter gives only the European Parliament the honour for the inclusion of the second sentence in Art. 2(1). Schröter relied on Deringer, '*Das Wettbewerbsrecht der Europäischen Wirtschaftsgemeinschaft*, *VO* No 26, art. 2, rn. 10 and 24 (Verlag Handelsblatt, 1962-1967) , where it is stated that a majority of the Member States wanted clarification of Art. 2, and that this clarification, in the form of Art. 2(1), second sentence, essentially copied § 100(1) of the German law on competition then in force.
49. See nn. 36 and 37 above for references to the discussion in the literature.
50. See Commission Decision 85/76/EEC of 7 Dec. 1984 relating to a proceeding under Art. 85 of the EEC Treaty (IV/28.930 – *Milchförderungsfonds*) (OJ L 35, 7 Feb. 1985, 35). For the same approach to the interpretation of Art. 2(1), second sentence, see Commission Decision

The Commission's narrow interpretation of Article 2(1), second sentence, does not appear entirely loyal, in the light of the negotiations in the Council which led to the adoption of the provision.[51] However the interpretation which the Commission settled on was allowed to stand for a number of years until 1992, when the Commission changed its view and acknowledged that Article 2(1), second sentence, is an independent exception.[52]

[c]   The Interpretation of the Court of Justice

In 1995 the question was finally settled by the decision in the *Dijkstra* case. The Court of Justice established that Article 2(1), second sentence, concerns an independent exception.[53] Dijkstra, who held some Dutch milk quotas, was excluded from a cooperative association by the association's general meeting. The reason for this was that he had refrained from delivering the whole of his milk production to the association, contrary to the articles of the association. In extension of the exclusion, the cooperative association claimed compensation from Dijkstra under the articles of association. Dijkstra responded by claiming before the Dutch courts that the articles of association were invalid, because they were contrary to Article 101(1) TFEU. Against this background the Dutch court referred the matter to the Court of Justice for a preliminary ruling on the interpretation of Article 2(1), second sentence.[54]

The Court stated, by way of introduction, that the interpretation of Article 2(1), second sentence, must have regard for the genesis of the provision and the purpose of the Regulation. According to the *travaux préparatoires* for the Regulation, the second sentence of Article 2(1) was not in the Commission's original draft for the Regulation, but it was included at the request of the European Parliament. The Court found that, in

---

88/491/EEC of 26 Jul. 1988 relating to a proceeding pursuant to Art. 85 of the EEC Treaty (IV/31.379 – *Bloemenveilingen Aalsmeer*) (OJ L 262, 22 Sep. 1988, 27). In point 151 of the Decision, the Commission stated that 'even if the special circumstances specified in the second sentence were to constitute a separate exception, this additional criterion would still not be satisfied'.

51. In support of its decision, the Commission referred to a footnote in the Court of Justice's judgment in Case 71/74 *Frubo* [1975] ECR 563, para. 8. In this case, the Court of Justice only cited the first sentence of Art. 2(1), and refrained from referring to the second sentence. However, this was in connection with the Court considering a formal claim that in the case in question the Commission had not made a decision about the application of the then Council Regulation No 26/62. It is this hardly possible to read anything into this judgment on the scope of Art. 2(1), second sentence.

52. See Commission Decision 92/444/EEC of 30 Jul. 1992 relating to a proceeding pursuant to Art. 85 of the EEC Treaty (IV/33.494 – Scottish Salmon Board) (OJ 1992 L, 27 Aug. 1992, 37).

53. Prior to this judgment, there was support for this view in Ries & Guida, 'L'application des règles de concurrence du traité CE à l'agriculture', CDE 60 (169–172) (1968); Ottervanger, 'Antitrust and agriculture in the common market', Fordham Corp. L. Inst. 203 (218) (1990); and Olmi in *Le droit de la CEE (Commentaire Megret)*, Vol. 2 – '*Politique agricole commune*' 280-282 (Université de Bruxelles, 2nd ed., 1991). See also Cooke, *Agricultural Law for the European Union* 67 (70–80) (Heusel & Collins eds., Academy of European Law and Irish Centre for European Law, 1999); and Mensching, *Agricultural Law for the European Union* 87 (Heusel & Collins eds., Academy of European Law and Irish Centre for European Law, 1999).

54. The *Dijkstra* case consisted of several Joined Cases, which all raised the same question of principle.

following the request of the Parliament, the Union legislator 'sought to introduce an exception applying in favour of agreements, decisions and practices of farmers' which otherwise fulfilled the requirements referred to. In extension of this, the Court of Justice stated that: 'that desire to protect agricultural cooperatives is apparent from the reasons given for the regulation, and in particular from the fourth recital in the preamble to Regulation No 26, which states that special attention is warranted in the case of farmers' organisations'.[55] The Court of Justice concluded that: 'To interpret the second sentence as having no independent meaning would run squarely counter to the wishes of the legislature'.[56]

Most recently, in the *Luttikhuis* case, the Court of Justice found that Article 2(1) contains three exceptions to the main rule that Article 101(1) TFEU applies to the agricultural sector. As for the third exception, the Court stated that this only applies if three cumulative conditions are satisfied. First, the farmers or cooperative associations in question must belong to a single Member State. Second, the agreements must not concern prices but rather the production or sale of agricultural products or the use of joint facilities for the storage, treatment or processing of agricultural products. And third, the agreements must not jeopardize the objectives of the common agricultural policy.[57]

[d] Conclusion on the Interpretation of Article 2(1), Second Sentence

The main aim of the exception in Article 2(1), second sentence, is to prevent ordinary farmers' associations being caught by Article 101(1) TFEU. Obligations arising from agreements, for example in the form of articles of association between farmers and their associations, will be covered by the exception if the conditions are otherwise fulfilled. The requirement in the second sentence, that there should be 'no obligation to charge identical prices', was presumably included in order to prevent the establishment of ordinary pricing cartels. For example, an agreement between beef cattle producers not to sell their cattle below a given minimum price would fall outside the exception and would thus be covered by the main rule in Article 101 TFEU, whereby agreements that restrict competition are prohibited. On the other hand, if the producers sell through an association and receive a proportionate share of the association's revenues, it must be assumed that this does not amount to an 'obligation to charge identical prices', so that the exception in the second sentence will apply. If sales through an association were to constitute a price agreement, it would be impossible in practice for any sales association to find protection under Article 2(1), second sentence.[58]

---

55. See n. 47 above, on the inclusion of the fourth recital in Council Regulation No 26/62.
56. See Joined Case C-319/93, C-40/94 and C-224/94 *Dijkstra* [1995] ECR I-4471, paras. 17–20. The judgment is a good example of the significance of *travaux préparatoires* for the interpretation of Union legislation.
57. Case C-399/93 *Oude Luttikhuis* [1995] ECR I-4515, paras. 24–27.
58. Bellamy & Child, *European Community Law of Competition* 179, n. 654 (Oxford University Press, 6th ed., Roth & Rose 2008).

The exception in Article 2(1), second sentence, only applies to agreements etc. between farmers and their associations or between such associations. The exception does not cover agreements between farmers and non-farmers, such as distributors or wholesalers.[59]

Article 2(1), second sentence, does not expressly exclude agreements etc. from the scope of Article 101(1), if 'the Commission finds that competition is thereby excluded or that the objectives of Article 39 of the Treaty are jeopardised'. In practice, this part of the provision is used to establish that the articles of association of a farmers' cooperative association that both obliges the members of the cooperative to deliver their products to the association and imposes an obligation on members to pay substantial compensation to the association on exiting the association or being excluded from it, can be contrary to the conditions referred to in Article 2(1), second sentence.

In the *Campina* case, the Commission approved some provisions in the articles of association of the Dutch dairy cooperative Campina, after these had been amended after the Commission had initiated an investigation into Campina's articles of association in the light of Article 101 TFEU. A Belgian dairy cooperative had complained to the Commission about a number of the provisions in Campina's articles of association. According to the provisions in question, the members were in principle bound to supply the whole of their milk production to Campina. If the members wanted to leave the cooperative, there was a notice period of one month, and they could only leave at the end of a financial year, having paid compensation on exit amounting to 10% of their average annual income from the sale of their milk to the cooperative.

The Commission did not approve the payment of compensation on exit amounting to 10% of their average annual milk sales revenue. This financial obligation was a significant economic hindrance for members wishing to exit the cooperative. The Commission found that the consequence was that the members were in practice bound to supply all their milk to the cooperative for an indefinite period. This not only restricted the economic freedom of Campina's members, but also the freedom of Campina's competitors to obtain the milk which Campina's members produced. The Commission concluded that the agreement on exclusive deliveries, combined with the obligation to pay compensation on exit amounting to 10% of their average annual milk sales revenue, was subject to Article 101(1) TFEU, that it could not be exempted pursuant to Article 101(3) TFEU, and that it was not covered by the exceptions for agricultural market organizations pursuant to what is now Regulation (EC) No 1184/2006. The provisions in the articles of association were not necessary for attaining the objectives set out in Article 39; on the contrary, they made it more difficult to attain the objectives.

On this basis, Campina amended its articles of association, so that the association's members could leave the association at three different time in the course of the year (1 April, 1 September and at the end of the financial year), without paying

---

59. Commission Decision 78/66/EEC of 2 Dec. 1977 relating to a proceeding under Art. 85 of the EEC Treaty (IV/28.948 – Cauliflowers), Part III, point 2. On the *Cauliflower* case, see above in §8.03[A][3].

compensation for exiting as long as they gave two years notice. The members could also exit with three months notice, but in this case they had to exit on 1 April, and with payment of compensation of 4% of their average annual income from the sale of their milk to the cooperative.

The Commission stated that the amendments to the articles of association meant that competition between Campina and other dairies for access to the milk produced by Campina's members would be strengthened, to the advantage of milk producers who would have greater commercial freedom, without this having negative effects on Campina's economic and financial stability. The obligation of exclusive supply, combined with the new articles of association on the members' right to exit the cooperative still had the effect of restricting competition, but the Commission found that the conditions could be approved, given the structure of the milk market and Campina's position on the market. Restrictions on competition in the form of exclusive delivery obligations for up to two years in favour of a cooperative company that does not have a dominant market position are thus covered by the exception in what is now Regulation (EC) No 1184/2006.[60]

In the *Luttikhuis* case, the Court of Justice stated that it could not rule out that the cumulative effect of clauses in statutes tying the members to an association for long periods, and thereby depriving them of the possibility of approaching competitors, could jeopardize one of the objectives of the common agricultural policy. The provisions in articles of association can be contrary to the objective of Article 39 TFEU of increasing the individual earnings of persons engaged in agriculture, since binding them to a cooperative means they cannot exploit the advantages of competition for purchase prices which processing companies pay for raw materials.[61]

Undertakings which claim that agreements which distort competition are necessary for the attainment of the objectives of Article 39 TFEU, must give evidence of this necessity. The *Suiker Unie* case concerned, among other things, a concerted practice between two undertakings in the sugar sector, *NV Centrale Suiker Maatschappij* (CSM) and *Raffinerie Tirlemontoise* (RT). CSM and RT had the practice of not competing in each other's markets, and RT supplied to CSM at higher prices. The undertakings argued that even though it must be assumed that this conduct constituted a concerted practice pursuant to Article 101, it was nevertheless lawful. They argued that their practice fell within the exception in Article 2(1), first sentence, of the Regulation, as a practice that was 'necessary for attainment of the objectives set out in Article 39 of the Treaty'. The Court of Justice rejected this claim. It ruled that there were no grounds for examining whether the mere payment of a higher price than the minimum price could 'ensure a fair standard of living for the agricultural community', in this case sugar beet producers, which is one of the aims stated in Article 39(1)(b) TFEU. The Court of Justice found that 'in any case CSM has not attempted to show with any degree of accuracy that only its purchases from RT enabled it to offer' a higher price to sugar beet

---

60. On the *Campina* case, see the XXIst report on competition policy 1991, European Commission, 75–76, points 83–84.
61. Case C-399/93 *Oude Luttikhuis* [1995] ECR I-4515, para. 28.

producers. In other words, the undertakings had not proved that their practice was necessary for the attainment of the objectives of the common agricultural policy.[62]

[B]   The Common Organization of Agricultural Markets (Regulation (EC) No 1234/2007 – the 'Single CMO Regulation')

As stated above in section §8.03, Articles 175–179 of the Single CMO Regulation lay down special provisions on the application of the TFEU competition rules to undertakings that produce or trade in certain agricultural products. Article 175 states that main rule that Articles 101–106 TFEU and their implementing provisions apply to all agreements, decisions and practices referred to in Articles 101(1) and 102 TFEU and which relate to the production of or trade in all the products that are subject to the Single CMO Regulation.[63] This main rule applies unless otherwise provided for in the Single CMO Regulation, including Articles 176 and 177a. Apart from the restriction as to the relevant products and the exception for the Regulation's special rules, Article 175 is identical to Article 1a of Regulation (EC) No 1184/2006.

Article 176 follows the basic structure known from Regulation (EC) No 1184/2006 as, apart from a few differences of language, it is identical to Article 2 of Regulation (EC) No 1184/2006. The interpretation of Article 2 of Regulation (EC) No 1184/2006 thus applies equally to the interpretation of Article 176 of the Single CMO Regulation.[64]

Articles 176a–179 contain special rules for particular agricultural sectors which mean that, despite the rule in Article 175, Article 101(1) does not apply to agreements, decisions and concerted practices of recognized inter-branch organizations.

[1]   The Fruit and Vegetables Sector

Article 176a(1) provides that Article 101(1) of the Treaty does not apply to the agreements, decisions and concerted practices of recognized inter-branch organizations in the fruit and vegetables sector which have the object of carrying out the activities referred to in Article 123(3)(c) of the Single CMO Regulation.[65]

According to Article 176a(2), this exemption only applies if two formal conditions are fulfilled. First, the agreements, decisions and concerted practices must have been notified to the Commission. Second, the Commission must not have declared, within two months of receipt of all the details required, that the agreements, decisions or concerted practices are incompatible with Union rules. Article 176a(3) expressly states

---

62. Joined Cases 40 to 48, 50, 54 to 56, 111, 113 and 114/73 *Suiker Unie* [1975] ECR 1663, paras 211–216. It is not sufficient to be able to refer to the fulfilment of one or more of the objectives of the common agricultural policy, all objectives must be satisfied; see §8.03[A][3] above. See Ch. 2, §2.05, on the interpretation of Art. 39 TFEU, which lays down the objectives of the common agricultural policy.
63. See Ch. 4, §4.02[B], on the scope of the Single CMO Regulation.
64. On the interpretation of Art. 2 of Regulation (EC) No 1184/2006, see §8.03[A][2]–§8.03[A][4] above.
65. See Ch. 4, §4.05[B], on private organizations and associations as a part of the Single CMO Regulation, including Art. 123.

that the agreements etc. may not be put into effect before the end of the two-month period.

Moreover, according to Article 176a(4), the exemption only applies if five substantive conditions for the agreement etc. in question are fulfilled. If the conditions are not fulfilled, the agreement etc. will be regarded as being incompatible with Union law. First, the agreement etc. must not lead to the partitioning in any form of markets within the Union. Next, the agreement etc. must not affect the sound operation of the market organization. The third condition is that the agreement etc. may not create distortions of competition which are not essential to achieving the objectives of the common agricultural policy pursued by the inter-branch organization activity. Fourth, the agreement etc. may not entail the fixing of prices, without prejudice to activities carried out by inter-branch organizations in the application of specific Union rules.[66] Finally, agreements etc. may not create discrimination or eliminate competition in respect of a substantial proportion of the products in question.

According to Article 176a(5), it is the Commission that monitors compliance with these conditions. If, following the expiry of the two-month period referred to above, the Commission finds that the conditions for applying the exemption have not been met, it shall take a Decision declaring that Article 101(1) TFEU applies to the agreement etc. The Commission cannot give such a Decision retrospective effect unless the inter-branch organization concerned has given incorrect information or abused the exemption provided for in Article 176a(1).[67]

Both the formal and substantive conditions for the exemption referred to above are very vaguely worded and leave a wide margin of discretion to the Commission.

Agreements that are exempt under the rule in Article 176a(1) typically extend over a number of years. For this reason, Article 176a(6) contains a rule stating that in the case of multiannual agreements, the inter-branch organization's notification for the first year is valid for the subsequent years of the agreement. However, in this case the Commission may, on its own initiative or at the request of 'another Member State', issue a finding at any time to the effect that the agreement no longer fulfils the conditions for exemption from the provisions in Article 101(1) TFEU. Article 176a(6) refers to 'the agreement', and not to 'agreements, decisions and concerted practices', as in the related provisions. It must be assumed that if a decision or concerted practice that extends over several years has been notified to the Commission in one year, then the notification also applies to subsequent years.

[2]   The Tobacco Sector

Article 177 of the Single CMO Regulation contains a provision that exempts agreements and concerted practices of recognized inter-branch organizations in the tobacco sector from the provisions in Article 101(1) TFEU. In its structure and content, the rule is

---

66. By comparison, in the tobacco sector in addition to prices, the fixing of quotas is also prohibited. See §8.03[B][2] below.
67. See Ch. 3, §3.02[D], on the fundamental principle of the protection of legitimate expectations.

virtually interchangeable with Article 176a on the fruit and vegetables sector. However, according to its wording, Article 177(1) does not exempt 'decisions', and the period within which the Commission must act is three months.[68] In the statement of the substantive conditions which the individual agreement or concerted practice must fulfil, Article 177(2) not only prohibits the fixing of prices, but also the fixing of quotas.

As a special rule, Article 178 gives recognized inter-branch organizations in the tobacco sector the right to request the Commission that, in the areas in which they operate, certain of their agreements or concerted practices be made binding for a limited period on individuals and groups in the sector concerned which are not members of the trade branches which they represent. The Commission's approval depends on the fulfilment of a number of conditions. Among other things, the rules for which extended application is requested must concern one or more expressly stated objectives, for example the definition of minimum qualities, the use of certified seed, and the monitoring of product quality.

[3]   The Milk and Milk Products Sector

Article 177a provides for exemptions from the provisions of Article 101(1) TFEU for agreements, decisions and concerted practices for recognized inter-branch organizations in the milk and milk products sector. The rule is structured in the same way as in the corresponding rules for the fruit and vegetables and tobacco sectors. For agreements etc. in the milk and milk products sector, it is a requirement that the recognized inter-branch organization should seek to carry out the activities referred to in Article 123(4)(c) of the Single CMO Regulation.[69] According to Article 177a(2), the deadline for the Commission's decision that the exemption should not apply is three months. Article 177a(7) contains special authority for the Commission to adopt implementing acts laying down measures necessary for the uniform application of Article 177a(7).[70]

§8.04   THE RULES ON STATE AID TO UNDERTAKINGS IN THE AGRICULTURAL SECTOR

Article 107(1) TFEU prohibits any aid granted by a Member State or through State resources in any form whatsoever which distorts or threatens to distort competition by

---

68. Article 177(1) refers to 'the aims referred to in Article 123(c) of this Regulation', without naming the para. of Art. 123 in question. In the comparable provision, Art. 176a(1) refers to Art. 123(3)(c). The lack of reference in Art. 177(1) is due to the fact that the provision is derived from the original version of the Single CMO Regulation in 2007, at a time when Art. 123 only had one paragraph. Today, the reference in Art. 177(1) should be understood as a reference to Art. 123(1)(c).
69. See above, in Ch. 4, §4.05[B], on private sector organizations and associations that are part of the Single CMO Regulation, including Art. 123.
70. In Art. 179, the Commission is given authority to lay down detailed implementing provisions for Arts 176a, 177 and 178, including rules on notification and publication. See Commission Regulation (EC) No 709/2008 laying down detailed rules for implementing Council Regulation (EC) No 1234/2007, as regards inter-branch organizations and agreements in the tobacco sector.

favouring certain undertakings or the production of certain goods, in so far as it affects trade between Member States. The Treaty concept of State aid is extensive, as it affects the measures of public bodies that, in one way or another, seek to favour certain undertakings in preference to others. It does not matter whether the public bodies are state, regional or local authorities. And the concept of aid is not restricted to subsidies paid directly to undertakings, but also includes indirect aid. In the case law of the Court of Justice, many national arrangements have been held to be indirect aid, contrary to the prohibition of State aid, for example favourable tax treatment in the form of tax reliefs or special depreciation allowances, or public guarantee or insurance arrangements, where the premiums paid do not cover the costs of the arrangement. However, it is a basic condition for aid to fall under the prohibition in Article 107(1) that the aid is provided directly or indirectly from State resources. If aid is provided as part of general economic policy, the arrangement will fall outside the scope of the State aid rules.[71]

Article 42 TFEU refers generally to the provisions of the chapter relating to rules on competition, and thus to the Treaty's State aid rules in Articles 107–109. As a special rule of agricultural law, Article 44 TFEU provides that under certain conditions a Member State may impose import duties on an agricultural product if the product has been produced with the help of State aid in the exporting Member State.

The Union legislator has laid down provisions on the application of the State aid rules in the agricultural sector in Council Regulation (EC) No 1184/2006 applying certain rules of competition to the production of, and trade in, agricultural products; in the Single CMO Regulation and in the Rural Development Regulation No 1698/2005. Even if a national aid measure may be allowed under the Treaty State aid rules, it is still subject to the rules on the free movement of goods.[72]

It follows from Article 42's reference to the State aid rules that the State aid rules only apply to the agricultural sector to the extent determined by the European Parliament and the Council. It is the Union legislator that decides whether, and if so to what extent, the State aid rules apply.

The Court of Justice gave a significant contribution to the interpretation of Article 42's reference to the State aid rules in annulment proceedings brought against the Council by the Commission. The Commission claimed annulment of two decisions of the Council. The decisions were adopted under the authority of what is now Article 108(2), third paragraph, and permitted the grant of extraordinary State aid by Italy and France for the distillation of some wine.[73]

In support of its claim for annulment of the Council Decisions, the Commission argued that the Council did not have the necessary powers to adopt the Decisions. It stated that, according to the wording of Article 108(2), third paragraph, TFEU the

---

71. On the Treaty rules on State aid in general, see Bellamy & Child, *European Community Law of Competition* 1497–1613 (Oxford University Press, 6th ed. Roth & Rose 2008).
72. See Ch. 3, §3.02[F], on Art. 34 TFEU and agricultural products in the internal market; and for a more detailed discussion of Regulation (EC) No 1184/2006 in connection with State aid, see §8.04[A] below.
73. Case C-122/94 *Commission v. Council* [1996] ECR I-881. At that time authority for the Council's decisions was given in Art. 93(2), third paragraph, of the EC Treaty.

Council can only derogate from the provisions in Article 107 or the regulations referred to in Article 109 TFEU, but not from other provisions of Union law. This interpretation of the wording of Article 108(2), third paragraph, was rejected by the Court of Justice. The Court stated that in fact Article 108(2), third paragraph, provides that, in derogation from the provisions of Article 107 or from the regulations provided for in Article 109, upon the application of a Member State, the Council may, acting unanimously, decide that aid which that State grants or intends to grant is to be considered compatible with the common market, if such a decision is justified by exceptional circumstances.[74]

Next, the Court referred to Article 42 TFEU. Under the authority of the market organization for wine then in force, the Council had the power to decide that Articles 107 and 108 applied to the production of and trade in wine and must. The Court thus found that the Council's powers pursuant to Article 108(2), third paragraph, also apply in the wine sector, within the limits indicated by that provision, namely the existence of exceptional circumstances.[75]

As another argument in support of its claim, the Commission stated that in adopting the Decisions, the Council had misused the procedure, as by referring to Article 108(2), third paragraph, the Council had derogated from the rules relating to the common organization of the market in wine. To the extent that the Decisions altered the market organization, the Council should have adopted amendments pursuant to Article 43 TFEU, on the adoption of legislation governing agriculture.

The Court of Justice also rejected this argument, merely stating that the reference in the market organization to Articles 107–109 TFEU did not contain any conditions additional to those laid down in the provisions referred to. The Commission's argument that the disputed aid measures interfered with the common organization of the market in wine should thus only be examined in more detail if the Commission could prove that the Council had exceeded the scope of its discretion under Article 108(2), third paragraph. On the basis of a careful examination of the Council's exercise of its discretion, the Court concluded that the Council had not been clearly wrong in the exercise of its discretion.[76]

[A]   Council Regulation (EC) No 1184/2006 Applying Certain Rules of Competition to the Production of, and Trade in, Agricultural Products

Article 3 of Council Regulation (EC) No 1184/2006 applying certain rules of competition to the production of, and trade in, agricultural products provides that Article 108(1) and (3), first sentence, apply to aid granted for production of, or trade in, the products that fall within the scope of the Regulation.[77] The restriction of the effect of the State aid rules pursuant to the Regulation is presumably due to the fact that the

---

74.   *Ibid.*, paras 8–9.
75.   *Ibid.*, paras 10–13.
76.   *Ibid.*, paras 14–15 and 25. See above in Ch. 3, §3.01-§3.02, on judicial review of the exercise of discretion given by the Union legislator.
77.   On the scope of Regulation (EC) No 1184/2006, see §8.02 above.

Regulation merely carries forward the old EEC Council Regulation No 26/62. This rule was presumably laid down because in 1962 it was originally expected that national aid measures for agriculture would only be of short duration, until the common market organizations covering the whole of agriculture were in place.[78]

If a State aid measure does not only concern matters falling within the scope of Regulation (EC) No 1184/2006, this presumably means that the general State aid rules apply to the whole of the arrangement. In proceedings before the Court of Justice for breach of the Treaty, the Commission argued that a French arrangement, which involved a 'solidarity grant' paid to the poorest farmers, and which thus did not only concern production of, or trade in, agricultural products, was in breach of the Treaty rules on State aid. The Court of Justice decided the case solely on the basis of the State aid rules in the Treaty.[79]

Article 108(1) TFEU states that, in cooperation with Member States, the Commission should keep under constant review all systems of aid existing in the Member States, and propose to the Member States any appropriate measures required by the progressive development or functioning of the internal market.

Article 108(3), first sentence, states that the Commission must be informed of any plans to grant or alter aid in sufficient time to enable it to submit its comments. Under Article 108(2) TFEU, the Commission can require a Member State to abolish or alter an aid measure, but since Article 3 of Regulation (EC) No 1184/2006 does not refer to Article 108(2) TFEU, the Commission cannot exercise this power if the aid measure relates to agricultural products that are subject to Regulation (EC) No 1184/2006. On this basis, Article 3 of Regulation (EC) No 1184/2006 means that even if the Member States are required to inform the Commission about aid measures, formally the Commission cannot do anything other than take note of the information.[80]

In the Community guidelines for State aid in the agriculture and forestry sector 2007–2013, the Commission has given a full account of its view on the possibilities for national aid schemes in agriculture.[81] The guidelines express a change to the common agricultural policy, including establishing a new framework for aid to rural development areas, as laid down in Council Regulation (EC) No 1698/2005 on support for rural development by the EAFRD.[82] The Commission's aim with the guidelines is to ensure compatibility between the various aid measures and the Union's obligations under the WTO Agreement on Agriculture.[83]

---

78. Bellamy & Child, *European Community Law of Competition* 1209 (Oxford University Press, 5th ed., 2001).
79. See Case 290/83 *Commission v. France* [1985] ECR 439. In his Opinion in the case, Advocate General Mancini also did not rely on the rules of agricultural law.
80. See Case 114/83 *Société d'initiatives et de coopération agricole* [1984] ECR 2589, paragraph 27.
81. Community guidelines for State aid in the agriculture and forestry sector 2007 to 2013 (OJ C 319, 27 Dec. 2006, 1). See also Information from the Commission – Community Guidelines for State aid in the agriculture sector (OJ C 28, 1 Feb. 2000, 2).
82. See above in Ch. 5 on Council Regulation (EC) No 1698/2005 on support for rural development.
83. See McMahon, *EU Agricultural Law* 17–18 (Oxford University Press, 2007).

[B]   The Single CMO Regulation (Council Regulation (EC) No 1234/2007)

Article 3 of Regulation (EC) No 1184/2006 is not alone in regulating the application of the State aid rules to agriculture. It is pointed out in the preamble to the Single CMO Regulation that the many regulations that have previously established common market organizations for various agricultural products have often contained provisions that have meant that the State aid provisions were applicable to State aid for the agricultural products in question.[84] Even where market organizations have not contained a provision on the application of the State aid rules, the existence of aid measures in the individual market organization has meant that the Member States have been prevented from establishing national aid measures, unless this was permitted in the market organization itself. In the *Pigs and Bacon Commission* case, the Court of Justice ruled that a Member State's recourse to the State aid provisions in Articles 107–109 TFEU 'cannot receive priority over the provisions of the Regulation on the organization of that sector of the market'.[85]

Article 180 of the Single CMO Regulation lays down the presumptive rule that, unless otherwise provided in the Regulation, Articles 107, 108 and 109 TFEU apply to the production of and trade in the products that are subject to the Regulation. Thus, as distinct from the presumptive rule in Regulation (EC) No 1184/2006, under the Single CMO Regulation the main rule is that all the Treaty State aid rules apply.

However, the first exception to this main rule is laid down already in Article 180, second paragraph. Articles 107–109 TFEU do not apply to a number of payments which Member States make pursuant to the Single CMO Regulation. This applies, for example, to extraordinary market aid measures in connection with animal diseases or natural disasters, aid for the supply of milk to pupils in schools and various other aid measures in the fruit and vegetables sector.[86]

As a second exception to the main rule, there is a special rule for milk and milk products in Article 181.[87] Article 181 contains an exception to the main rule in the sense that the provision goes even further than the main rule in laying down an absolute prohibition of State aid. The first paragraph of Article 181 states that aids, the amount of which is fixed on the basis of the price or quantity of products, is prohibited. However, as an exception to the exception, the provision does not cover aid that is covered by Article 107(2) TFEU.[88] Article 181, second paragraph, states that there is an absolute prohibition of national measures permitting equalization between the prices for milk and milk products. There is no exception to this prohibition.

---

84. See recital 83 in the Single CMO Regulation.
85. Case 177/78 *Pigs and Bacon Commission* [1979] ECR 2161, para. 11.
86. Article 180, second paragraph, of the Single CMO Regulation refers to payments made under Arts 44 to 48, 102, 102a, 103, 103a, 103b, 103e, 103ga, 104, 105, 182 and 182a, Subsection III of section IVa of Ch. III of Title I of Part II and section IVb of Ch. IV of Title I of Part II of the Regulation by Member States in conformity with the Regulation. With regard to Art. 103n(4) only Art. 88 of the Treaty does not apply. See Ch. 4, §4.04[D], on the various aid arrangements in connection with the Single CMO Regulation.
87. Article 181 specifies the group of milk and milk products by reference to the products listed in Part XVI of Annex I.
88. Article 107(2) TFEU lists some forms of aid that are compatible with the internal market.

The third exception to the main rule applies to State aid for wine. Article 182a provides that Member States may grant national aid to wine producers for the voluntary or mandatory distillation of wine in justified cases of crisis.[89] In order for such an aid measure to be lawful, the individual Member State must comply with some formal and substantive conditions. There is a formal requirement that a Member State that wishes to grant such aid must submit a duly substantiated notification to the Commission. According to Article 182a(4), the Commission decides whether the measure is approved and the aid may be granted. As a consequence, even though Article 182a applies in the event of a crisis, and there is thus often a pressing need for quick action, Member States cannot initiate their aid measures before they have received the approval of the Commission. The provision does not set a deadline by when the Commission must give its approval. This means that, according to the general rules of administrative law, the Commission's decision must be given within a reasonable period, and if the Commission has not decided within a reasonable period, a Member State may lawfully initiate its aid measures.[90]

A condition for obtaining the approval of the Commission pursuant to Article 182a(2) is that the proposed aid should be proportionate and should make it possible to resolve the crisis. This provision expresses the general principle of proportionality in Union law and it should thus be interpreted in line with this principle.[91] The second substantive requirement for a proposed aid measure, in Article 182a(3), is that the overall amount of aid available in a Member State in any given year may not exceed 15% of the globally available funds per Member State laid down in Annex Xb for that year.[92] Finally, Article 182a(5) imposes the requirement that the alcohol resulting from crisis distillation must be used exclusively for industrial or energy purposes. The purpose of this requirement is to avoid distortion of competition, though there is no reference to the nature of this competition.

The last exception to the main rule in Article 180 of the Single CMO Regulation, that the State aid rules apply to the products that are subject to the Regulation, is in Article 180 of the Regulation. This Article appears under the heading 'Specific national provisions', and gives rules for aid measures for specific Member States and agricultural sectors.

[C]   The Rural Development Regulation (Council Regulation (EC) No 1698/2005)

Article 88 of the Rural Development Regulation lays down special rules for the application of the Treaty State aid rules to aid that a Member State pays out for rural

---

89. According to Art. 182a(1), the right to provide aid subject to Art. 182a applies from 1 Aug. 2012.
90. Compare with Arts 176a, 177 and 177a of the Single CMO Regulation, on notification to the Commission of agreements, decisions and concerted practices, in which there are rules setting deadlines for the Commission to decide (two or three months), and the consequences of the Commission not deciding before the expiry of the deadline. On this, see §8.03[B] above. Article 182a(6) gives the Commission powers to adopt detailed rules for the application of Art. 182a.
91. See Ch. 3, §3.02[C], on the proportionality principle.
92. See Ch. 4, §4.04[D][8], on aid measures for wine and on Annex Xb to the Single CMO Regulation.

development.[93] The main rule is stated in Article 88(1), first paragraph, which states that Articles 107, 108 and 109 TFEU apply to support for rural development by Member States. As in the case of the main rule in the Single CMO Regulation, there are a number of exceptions to this main rule in the Rural Development Regulation. There are two main exceptions, and a group of exceptions for special forms of aid.

The first main exception, in Article 88(1), second paragraph, is that Articles 107, 108 and 109 TFEU do not apply to financial contributions provided by Member States in accordance with or pursuant to the Rural Development Regulation. In other words, State aid paid as part of rural development policy or as a supplement to rural development policy is wholly exempt from the State aid rules. It is expressly stated in Article 88(1), second paragraph, that this applies regardless of Article 89.[94]

The second main exception is in Article 89, according to which State aid intended to provide additional financing for rural development for which Union support is granted, must be notified to the Commission by Member States and approved by the Commission. The provision does not contain a deadline for approval or say whether initiating an aid measure requires prior approval.[95] A Member State's additional financing for rural development must be part of its rural development programme in accordance with Article 16 of the Rural Development Regulation, and approval of such additional financing is part of the approval of the Member State's overall rural development programme.[96] It is emphasized in Article 89 that this notification arrangement means that Article 108(3), first sentence, TFEU does not apply to rural development aid.

Article 88(2) to (6) of the Rural Development Regulation contains a large group of exceptions to the main rule in Article 88(1), first paragraph, that the Treaty State aid rules apply to rural development aid. The exceptions in this group are characterized by the fact that they concern special forms of aid and that they contain an absolute prohibition of a specific form of aid that is nevertheless lawful subject to certain conditions.

As the first special exception in this category, Article 88(2) provides that aid for the modernization of agricultural holdings which exceeds the percentages laid down in Annex I is prohibited.[97] However, this prohibition does not apply to three different forms of investment aid. The first of these is investment related to the conservation of traditional landscapes shaped by agricultural and forestry activities or to the relocation of farm buildings. However, these investments must be undertaken predominantly in the public interest. The second form of investment concerns the protection and improvement of the environment. And the third form of investment concerns improvements of the hygiene conditions of livestock undertakings, animal welfare and occupational safety in the workplace.

---

93. For further on structural and rural development policy, see Ch. 5.
94. On Art. 89, see immediately below.
95. See §8.04[B] above, on the corresponding question in connection with Art. 182a(4) in the Single CMO Regulation.
96. On national rural development programmes, see Ch. 5, §5.02[B], above.
97. On aid for the modernization of agricultural holdings, see Ch. 5, §5.02[A][1], above.

The second special exception, in Article 88(3), concerns State aid granted to farmers to compensate for natural handicaps in mountain areas and in other areas with handicaps. This form of aid is prohibited unless it fulfils the conditions in Article 37 of the Rural Development Regulation.[98] Moreover, even if additional aid exceeds the amounts fixed according to Article 37(3), the prohibition does not apply 'in duly justified cases'.

Under Article 88(4), as the third special exception State aid to support farmers who enter into agri-environmental or animal welfare commitments, but which fails to satisfy the conditions laid down in Articles 39 and 40, is prohibited. However, additional aid exceeding the maximum amounts set out in the Annex, pursuant to Articles 39(4) and 40(3), may be granted 'if duly justified'. There is express authority to derogate from the minimum duration of such aid commitments laid down in Articles 39(3) and 40(2).

As the fourth special exception, Article 88(5) provides that State aid to support farmers adjust to stricter standards based on Union legislation on environmental protection, public health, animal and plant health, animal welfare and occupational safety is prohibited. However, this does not apply if the aid in question satisfies the conditions laid down in Article 31.[99] However, even if the conditions of Article 31 are not fulfilled, additional aid exceeding the maximum amounts fixed pursuant to Article 31 may be granted to help farmers comply with national legislation which exceeds Union standards.

As the last special exception, to the main rule that the Treaty State aid rules apply to rural development, Article 88(6) provides that State aid to support farmers adapt to stricter standards based on national legislation on the environment, public health, animal and plant health, animal welfare and occupational safety is prohibited. However, this prohibition does not apply if otherwise provided for in Union legislation, or if the aid satisfies the conditions laid down in Article 31. Where justified under Article 31, additional aid exceeding the maximum amounts laid down under Article 31(2) and set out in the Annex to the Rural Development Regulation, may be granted.

---

98. See Ch. 5, §5.02[A][2], above on rural development aid for the improvement of the environment and landscape.
99. See Ch. 5, §5.02[A][1], on aid measures for the improvement of the competitiveness of agriculture and forestry

CHAPTER 9
International Relations

§9.01 THE EUROPEAN UNION IN AN INTERNATIONAL CONTEXT: EU, GATT AND THE WTO

The European Union (EU) plays a major international political role in the area of agriculture. The most important international negotiations on agriculture have taken place in the context of GATT and the WTO. Originally the EU was not a signatory to the GATT Agreement or a member of the WTO. The reason for this was that, under its own rules, until the Lisbon Treaty entered into force the EU could not have legal personality in the international arena. The European Community negotiated on behalf of its Member States in the GATT context until the setting up of the WTO in 1995, and it became a member of the organization in that year. The GATT Agreement and the WTO set significant limits to the Union's agricultural policy. These limits apply not only to the Union's rules governing imports and exports of agricultural goods, but also to domestic arrangements (i.e., within the Union).[1]

§9.02 THE DEVELOPMENT OF GATT AND THE WTO

As stated, GATT and in particular the WTO have had a significant influence on the development of the Union's agricultural policy. The obligations of the Union and its Member States under the GATT Agreement and the WTO set limits to the content of the Union's common agricultural policy. In particular, the WTO Agreement on Agriculture has pushed the development of the Union agricultural policy from being mainly based on support for products, primarily in the form of market interventions, to become support for producers, among other things farm subsidies based on area payments and set aside. This development has been driven by a desire to change the forms of agricultural support that are subject to commitments to make reductions under the

---

1. See §9.06 below on the significance of WTO law for Union law.

WTO rules to forms of support that are not subject to such commitments, and to satisfy the reduction requirements in this way.[2]

[A]   GATT and the International Trade Organization

During World War II, the idea arose of establishing a new economic world order. The idea was that the new economic order should be secured through three international organizations: the International Monetary Fund (IMF), the International Bank for Reconstruction and Development (IBRD – commonly known as the World Bank), and the International Trade Organization (ITO). The first two organizations were quickly established, while the negotiations on setting up the ITO were only partially successful. The text for the Charter to establish its organization was agreed in 1948, but the USA declined to ratify it, and so the organization did not come into being.[3]

Alongside the negotiations on the establishment of a trade organization, a number of counties were negotiating on an agreement on customs tariffs and trade, with the Western European countries setting the tone. These negotiations were soon successful, and in 1947 they resulted in the GATT Agreement. Without creating an international organization, the GATT Agreement contained two main groups of rules: on the one hand rules obliging contracting states to lower their customs tariffs, and on the other hand rules governing trade in goods between the contracting states.

When it became clear at the start of the 1950s that the ITO would be stillborn, the interested parties had to be content with the GATT Agreement.[4] However, the negotiations on the ITO were not wholly fruitless, as the GATT Agreement added a chapter on trade policy from the ITO draft Charter.

After this, GATT came to be the framework for the development of rules on international trade up until 1995. The development of these rules in the form of provisions in the Agreement itself and in special agreements linked to the GATT Agreement took place at negotiating rounds. These rounds were held at intervals of some years, and they are generally named after the place where the negotiating rounds

---

2.   See Usher, *EC Agricultural Law* 69, 78 (Oxford University Press, 2d ed., 2001). At p. 78, Usher states that 'there has been a considerable interaction between the terms of [the WTO Agreement on Agriculture] and the development of the EC's modern common organizations of agricultural markets'.
3.   In December 1950, US President Truman announced that, following several failed attempts, he would no longer try to get the Senate's approval of the agreement that would set up the International Trade Organization; see Diebold, *The end of the ITO* 1 (Princeton, N.J., 1952).
4.   While waiting for the negotiations on the International Trade Organization to succeed, the GATT Agreement provisionally entered into force, with some reservations, on 1 January 1948 by a protocol to the Agreement; see Protocol of Provisional Application of the General Agreement on Tariffs and Trade, UNTS 1950, Vol. 55, No 814, p. 308. See the GATT Agreement itself, UNTS 1950, Vol. 55, No 814, p.187. This provisional entry into force lasted until 1995.

started.[5] The negotiating rounds were not initiated according to a pre-determined plan, and the measures adopted depended on the political situation.[6]

[B] From the Dillon Round in 1960 to the Tokyo Round in 1973

The significant role that was beginning to be played in the context of GATT by the EEC was one of the causes of the initiation of the Dillon Round in 1960. Hitherto, the GATT system had been developed around industrial products, and in 1960 the negotiations particularly concerned Article XXIV on customs unions. While the Dillon Round led to the EEC lowering its customs tariffs on fruit and vegetables, it was first at the Kennedy Round in 1964 that there was the necessary political will to get to grips with the fundamental issue of international trade in agricultural products.

One of the goals of the Kennedy Round was to ensure 'acceptable conditions of access to world markets for agricultural products'.[7] The EEC put forward a plan to develop international trade in agricultural products, but this was received with some scepticism by the other countries and did not lead to any result. The countries outside the EEC regarded the proposal as an attempt by the EEC to protect its common agricultural policy which was in the process of being implemented.[8]

As the EEC's common agricultural policy created surplus production in Europe, its agricultural exports grew through the 1960s, often with special export subsidies. In 1973, the Tokyo Round was opened with the declared intention to create, 'as regards agriculture an approach to negotiations which, while in line with the general objectives of the negotiations, should take account of the special characteristics and problems in this sector'.[9] These negotiations too did not lead to significant results, which can be ascribed to the limited negotiating mandate given to the Commission by the Council.[10] The principles and mechanisms that formed the basis for the common agricultural

---

5. As an exception to this naming practice, the negotiating round in 1960–62 was named the Dillon Round after the then US Secretary of the Treasury, and the round in 1964–67, named the Kennedy Round after the former US President.
6. The negotiating rounds took place under the authority of Art. XXV(1), first sentence, of the GATT Agreement, which stated: 'Representatives of the contracting parties shall meet from time to time for the purpose of giving effect to those provisions of this Agreement which involve joint action and, generally, with a view to facilitating the operation and furthering the objectives of this Agreement.'
7. BISD (GATT Basic Instruments and Selected Documents) 1964, supplement 12, p. 36 (48, point 7).
8. On the EEC's proposal at the Kennedy Round see Warley, *International Economic Relations of the Western World 1959-1971* 287 (382–384) (Schonfield ed., Oxford University Press, 1976); Evans, *The Kennedy Round in American Trade Policy* 210–212 (Harvard University Press, 1971); and McMahon, *EU Agricultural Law* 79 (Oxford University Press, 2007).
9. BISD 1974, supplement 20, p. 19 (21).
10. According to Art. 207(3) of the TFEU, corresponding to Art. 133(3) of the Treaty Establishing the European Community (the EC Treaty), it is the Commission that conducts negotiations with one or more states or international organisations on issues of trade policy. The Commission negotiates on the basis of authorization from the Council.

policy could not be touched.[11] However, agreements were achieved in some specific areas, such as for dairy products and beef.[12]

[C]   The Uruguay Round and the Setting Up of the WTO

In 1994, the GATT states concluded a negotiating round that had started in Uruguay. The negotiating results of Uruguay Round were gathered together in a treaty which led to the setting up of the World Trading Organization (WTO). This did not mean that the slate was wiped clean and everything had to start again from the beginning.[13] The results of the GATT cooperation have been carried forward within the framework of the WTO. The Treaty which concluded the Uruguay Round included the Agreement Establishing the WTO and its Annex. Annex 1A is referred to as the GATT 1994 Agreement and consists of four parts. Three of these parts are: first, the GATT 1947 Agreement; second, the legal instruments adopted in connection with the original GATT Agreement; and third, six interpretations of certain provisions in the original GATT Agreement. Article II of the WTO Agreement states that the agreements and associated legal instruments included in the Annexes are integral parts of the WTO Agreement.

In the area of agriculture, the Uruguay Round produced a significant result in the form of the Agreement on Agriculture. The third paragraph of the preamble to the Agreement on Agriculture states that the objective of the Agreement is to produce 'substantial progressive reductions in agricultural support and protection sustained over an agreed period of time, resulting in correcting and preventing restrictions and distortions in world agricultural markets'.[14]

---

11. See Harris, *EEC Trade Relations with the USA in Agricultural Products* 6–9 (Wye College, 1977); and McMahon, *EU Agricultural Law* 79 (Oxford University Press, 2007). The Council's negotiating mandate for the Commission is reproduced in Harris, op. cit.,39.
12. For the agreement on beef see BISD 1980, supplement 26, 84; and for the agreement on dairy products see BISD 1980, supplement 26, 91.
13. The Treaty which concluded the Uruguay Round and led to the establishment of the WTO entered into force on 1 January 1995.
14. See the WTO Agreement on Agriculture preamble, third paragraph. On the Agreement, see below in §9.03[B], §9.04[B] and §9.05[B]. See Filipek, 'Agriculture in a World of Comparative Advantage: The Prospects for Farm Trade Liberalization in the Uruguay Round of GATT Negotiations', Harv. Intl. L.J. 123 (1989), on the complex negotiations of the Uruguay Round; and Cardwell, *The European Model of Agriculture* 320, n. 7 (Oxford University Press, 2004), for references to the comprehensive literature on the WTO Agreement on Agriculture.

§9.03 THE WTO RULES ON DOMESTIC SUPPORT MEASURES FOR AGRICULTURE

[A] The GATT Principles of Most-Favoured-Nation Treatment and National Treatment: GATT Articles I and III

One of the fundamental GATT rules is in Article I of the GATT Agreement, establishing the principle of most-favoured-nation treatment. The principle states that one contracting party must treat the goods from another contracting party on an equal footing as goods from any other contracting party.

The principle of most-favoured-nation treatment is supplemented by another important rule in Article III of the GATT Agreement, where the principle of national treatment prohibits discrimination against foreign goods. This principle states that once foreign goods have entered the territory of a country following customs clearance, the goods must be treated in the same way as domestic goods. However, the Article contains an exception in paragraph 8(b), which provides that 'the payment of subsidies exclusively to domestic producers, including... subsidies effected through governmental purchases of domestic products' are not contrary to the principle of national treatment.

In a dispute between the USA and the EEC on a support scheme for oilseed, a GATT Panel found that a subsidy which was not paid directly to the producers was not paid 'exclusively' to producers within the meaning of Article II(8)(b).[15] The Panel also found that the EEC's subsidy scheme did not ensure that the payments to producers were based on the prices which the processing undertakings had in fact paid the producers for the oilseed.[16]

[B] The WTO Agreement on Agriculture

[1] The Main Rule on the Commitment to Reduce Domestic Support Schemes

Article 3(2) of the WTO Agreement on Agriculture states that Members may not provide support in favour of domestic producers in excess of specified levels. There are exceptions to this rule in Article 6.

The levels referred to are laid down for each WTO Member in that Member's 'Schedule of concessions', which is either linked to the Marrakesh Protocol to the GATT 1994, which is the original GATT Agreement with amendments, or to an accession protocol. This protocol serves the important function of linking the then Members' schedules, which are part of the protocol, to the current GATT Agreement.

---

15. The case concerned aid paid on the basis of Regulation No 136/66/EEC on the establishment of a common organization of the market in oils and fats (OJ English special edition: Series I Ch. 1965-1966 p. 221). On the dispute settlement system of GATT and later the WTO, see Garcia Bercero, 'Trade Laws, GATT and the Management of Trade Disputes between the US and the EEC', Y.B. European L. 149 (1985) (163–173 on disputes under agricultural law).
16. BISD 1991, supplement 37, p. 86 (124–125, points 136–140).

The content of the individual Member's schedules can change over time, depending on the outcomes of the continuing WTO negotiations. Each Member's schedule consists of four parts, of which Part IV 'Specific commitments on domestic support and export subsidies on agricultural products' deals with the Member's obligations in the area of agriculture. With certain exceptions, the Union as a whole was obliged to reduce its support for producers by 20%, compared with the support paid during the period 1986–1988.

Further to Article 3(2), Article 6(1) of the WTO Agreement on Agriculture provides that the commitment to reduce support measures by each WTO Member pursuant to Part IV of its Schedule of concessions applies to all domestic support measures for agricultural producers, with the exception of domestic measures that are not subject to reduction pursuant to the criteria set out in this Article 6(2) to (5) and in Annex 2 to the Agreement on Agriculture.[17]

[2]   Exceptions to the Obligation to Reduce Domestic Support Payments

As a special exception, Article 6(5) provides that direct payments under production-limiting programmes are not subject to the Member's commitment to reduce domestic support. However, this exception is subject to certain conditions.

Annex 2 to the WTO Agreement on Agriculture pays down detailed rules on which domestic support schemes are not subject to the Member's commitment to reduce domestic support. Paragraph 1 in the Annex provides that in order for a support measure to be exempt from a Member's reduction commitments, the support measure must have 'no, or at most minimal, trade-distorting effects or effects on production'. It is also stated that this requires two conditions to be met. First, the support in question must be provided through a publicly-funded government programme not involving transfers from consumers. Second, the support must not have the effect of providing price support to producers.

With regard to support in the form of direct payments to producers, paragraph 2 of Annex 2 provides that, in order not to be subject to the Member's commitment to reduce domestic support, in addition to the condition in paragraph 1, a scheme must fulfil the further conditions in paragraphs 6 to 13. Paragraphs 6 to 13 lay down conditions linked to various forms of direct payments.[18]

Paragraph 6 concerns decoupled income support. In the WTO context, 'decoupled income support' means support that is calculated on the basis of the production in a given base year, and where the support does not depend on the actual production.[19] There are a number of conditions for such decoupled support to be excluded from the individual Member's obligations to reduce support schemes. Among

---

17. For a comprehensive discussion of Art. 6 of the WTO Agreement on Agriculture and the case law associated with it, see McMahon, *EU Agricultural Law* 127–139 (Oxford University Press, 2007).
18. Paragraph 5 contains a special rule for payments that are not covered by paras 6–13.
19. On 'decoupled support' in the common agricultural policy, see above in Ch. 1, §1.01, and above in Ch. 6.

other things, such payments must be made on the basis of clearly defined criteria, such as the income, status as a producer or landowner, factor used or production level in a defined and fixed base period.[20]

Rules on schemes whereby public bodies give income support or guarantees are found in Annex 2, paragraph 7. In order for such schemes to be excluded from a WTO Member's commitment to reduce domestic support, the scheme must comply with a number of conditions. Among other things, the scheme must be applicable to all producers and it may cover up to 70% of the individual producer's loss of income in the year in which they are entitled to income support.

It follows from paragraph 8 that the exemption from the Members' obligation to reduce their support for agriculture relates to support connected with natural disasters. Several conditions are made for such payments. One of the conditions is that losses of income, livestock (including payments for the veterinary treatment of animals), land or other production factors due to the natural disaster in question. According to paragraph 8, 'natural disasters' include outbreaks of diseases, pest infestations, nuclear accidents, and war on the territory of the Member concerned. Stocks of poultry, pigs or cattle that are slaughtered in connection of outbreaks of infectious diseases such as Newcastle disease, Teschen disease or foot-and-mouth disease, are 'natural disasters' within the meaning of paragraph 8. If a producer receives support pursuant to both paragraph 7 (income support or guarantee) and paragraph 8, the amount paid must be less than 100% of the producer's total loss.[21]

Paragraph 9 exempts structural adjustment assistance in the form of early producer-retirement programmes in the agricultural sector. As with the other exemptions from the obligations of WTO Members to reduce their support for agriculture, certain conditions are attached to this. Eligibility for such payments must be according to clearly defined criteria in programmes designed to facilitate the retirement of persons engaged in marketable agricultural production, or their movement to non-agricultural activities.[22]

While paragraph 9 concerns the retirement of people from agriculture, paragraph 10 concerns the removal of other resources from agricultural production (resource retirement). This paragraph exempts support schemes such as removing land or other resources, including livestock, from marketable agricultural production. In order to be

---

20. Usher, *EC Agricultural Law* 66 (Oxford University Press, 2001), argues that Council Regulation (EC) No 1251/1999 establishing a support system for producers of certain arable crops, then in force, was in accordance with para. 6. According to Art. 4(1) of the Regulation, area payments were calculated by multiplying the basic amount per tonne by the average cereal yield determined in the regionalization plan for the region concerned.

21. There are Union rules on support in connection with natural disasters in the rules of the Single CMO Regulation for the wine sector; see Ch. 4, §4.04[D][8], and in the Rural Development Regulation rules for axis 1 on support schemes for improving the competitiveness of agriculture and forestry and axis 2 on support schemes for improving the environment and landscape; see Ch. 5, §5.02[A][1] and §5.02[A][2].

22. Article 23 of Council Regulation (EC) No 1698/2005 (the Rural Development Regulation) contains rules on support for early retirement in the agricultural sector; see Ch. 5, §5.02[A][1], and this arrangement lies within the framework permitted in para. 9 of the WTO Agreement on Agriculture.

exempt pursuant to paragraph 10, the land must be set aside for at least three years and the livestock must be disposed of by slaughter 'or definitive permanent disposal'.

Support for investment in agriculture is under paragraph 11 exempt from the obligations of WTO Members to reduce their support for agriculture. This relates primarily to programmes designed to assist the financial or physical restructuring of a producer's operations in response to objectively demonstrated structural disadvantages. Investment support can also be for clearly defined government programmes for the reprivatization of agricultural land. The most important condition in paragraph 11 is in subparagraph (f), according to which payments must be limited to the amount required to compensate for the structural disadvantage.

Paragraph 12 exempts support paid by WTO Members for environmental programmes from their commitments to reduce support for agriculture. The most important condition for such programmes is that the amount of payments must be limited to the extra costs or loss of income involved in complying with the environmental programme.[23]

Support for geographically disadvantaged regions is exempted by paragraph 13 from WTO Members' commitments to reduce their support for agriculture. It is a condition that each disadvantaged region must be a clearly designated on the basis of neutral and objective criteria laid down in legislation, and the region's difficulties must be of a more than temporary nature.[24]

§9.04   THE WTO RULES ON IMPORT DUTIES AND QUANTITATIVE IMPORT RESTRICTIONS

[A]   The GATT Agreement's Prohibition of Quantitative Restrictions on Imports: Article XI

Article XI of the GATT Agreement contains a prohibition of quantitative restrictions on imports and exports. The first sentence of the provision states the main rule that neither imports nor exports may be subject to prohibitions or restrictions, whether by means of quotas, import or export licenses or other measures. Article XI(1) does not prohibit duties, taxes or other charges.

As a special exception to this main rule, Article XI(2) provides that import restrictions on agricultural and fisheries products are permitted under certain circumstances.[25] First, restrictions on imports are permitted where they are necessary to implement measures restricting the quantities of the like domestic product permitted to be marketed or produced.[26] Next, imports may be restricted where this is necessary to

---

23. The axes of the Rural Development Regulation contain rules on support for improvement of the environment and landscape; see Ch. 5, §5.02[A][2].
24. Axis 2 of the Rural Development Regulation deals with special support measures for farmers in areas with geographical disadvantages, such as hill farming see Ch. 5, §5.02[A][2].
25. For further on the exceptions in Art. XI(2) see Desta, *The Law of International Trade in Agricultural Products* 35–61 (Kluwer, 2002).
26. Article XI(2)(c)(i) of the GATT Agreement. A special proportionality condition is attached to this exception, since any restrictions applied under (i) must not be such as will reduce the total

remove a temporary surplus of the like domestic product by making the surplus available to certain groups of domestic consumers free of charge or at prices below the current market level.[27] Third, import restrictions may be imposed pursuant to Article XI(2) if this is necessary to restrict the quantities permitted to be produced of any animal product which is wholly or mainly directly dependent on the imported commodity. This is on condition that the domestic production of that commodity is relatively negligible.[28]

The exceptions relating to agricultural products in Article XI(2) effectively mean that the Union can justify import restrictions where these are linked to the regulation of the internal market. The Single CMO Regulation contains a number of rules regulating the production quantities of various agricultural products by setting quotas or establishing different forms of market intervention.[29]

However, the Union has not been able to totally avoid the prohibition of quantitative restrictions in Article XI. In the *Tomato concentrate* case, the USA complained about the Union's provisions on the import of processed fruit and vegetables. The Union had imposed a minimum price for tomato-concentrate that was imported into the Union. In addition, it was provided that upon the import of tomato-concentrate the importer had to pay a deposit to the Union's authorities, and the deposit would be lost if it was shown that the import was made at below the fixed minimum price. The USA argued that both the fixed minimum import price and the requirement to provide a deposit were contrary to Article XI(1) of the GATT Agreement.[30] The Union admitted this, but claimed that it was covered by the exceptions in Article XI(2)(c)(i) and (ii), according to which restrictions are allowed if they are part of measures to restrict the quantity of corresponding domestic products.[31]

The GATT Panel which decided the dispute did not accept the Union's argument. The Panel found that the Union measure was contrary to Article XI(1) because there was no Union restriction on the domestic production of the primary (fresh) agricultural product which would be made ineffective by the import of the processed product. The Panel also found that the Union could not rely on the exceptions in Article XI(2)(c)(i) and (ii).[32]

---

of imports relative to the total of domestic production, as compared with the proportion which might reasonably be expected to rule between the two in the absence of restrictions; see Art. XI(2), second paragraph, second sentence. See Usher, *EC Agricultural Law* 71 (Oxford University Press, 2d ed., 2001), whose presentation suggests that the requirement for proportionality applies to all the exceptions in Art. XI(2).

27. Article XI(2)(c)(ii) of the GATT Agreement.
28. Article XI(2)(c)(iii) of the GATT Agreement.
29. On the rules of the Single CMO Regulation regulating production, see Ch. 4, §4.03, and for example the milk quota scheme in §4.04[C][2] of Ch. 4.
30. BISD 1979, supplement 25, p. 68 (74).
31. *Ibid.*, at 75.
32. *Ibid.*, at 98–103. The case also raised questions in relation to Arts II and VIII. For a criticism of the Panel's report, see McMahon, *EU Agricultural Law* 75 (Oxford University Press, 2007) .

[B]   The WTO Agreement on Agriculture

In relation to import restrictions, the most significant contribution of the Agreement on Agriculture to free-trade in agricultural products is that WTO Members are required to convert their import restrictions into customs duties. The aim of this conversion is to ensure greater transparency and it is linked to requiring Members to reduce their import tariffs. Article 4(2) of the Agreement lays down the main rule that Members may not maintain, resort to, or revert to any measures which they have been required to convert into ordinary customs duties. The measures which must be converted into customs duties primarily include quantitative import restrictions, variable import levies, minimum import prices and discretionary import licensing.[33]

The conversion of variable import levies (an import restriction) to a fixed import duty (customs) means, for example, that the Union abandoned the system of indicative prices in the market organization for cereals, on the basis of which the variable import levy was calculated and which was aimed at the ongoing regulation of imports of cereals to the Union.[34]

§9.05   WTO RULES ON SUPPORT FOR EXPORTS

[A]   Article XVI of the GATT Agreement

Article XVI of the GATT Agreement governs export subsidies. Article XVI(3) binds the contracting parties to seek to avoid the use of subsidies on the export of primary products. However, if a contracting party grants directly or indirectly any form of subsidy which operates to increase the export of any primary product from its territory, such subsidy must not be applied in a manner which results in that contracting party having more than an equitable share of world export trade in that product, taking into account the shares of the contracting parties in trade in the product during a previous representative period, and any special factors which may have affected or may be affecting trade in the product. The interpretation of the wording of the provision has been the subject of some debate.[35]

In the Tokyo Round, the contracting parties made an Agreement on Interpretation and Application of Articles VI, XVI and XXIII of the GATT. Article 10(2) of this interpretative Agreement explains the term 'equitable share'. According to this, it means any case in which 'the effect of an export subsidy ... is to displace the exports of another signatory bearing in mind developments on world markets'.[36] However, this

---

33. See the footnote to Art. 4(2) of the Agreement on Agriculture. On the practice in relation to Art. 4(2), see McMahon, *EU Agricultural Law*, 108 (Oxford University Press, 2007).
34. See Council Regulation (EC) No 1528/95 amending Regulation (EEC) No 1766/92 on the common organization of the market in cereals.
35. The wording was included in the GATT Agreement of 1955, and already in 1959 its interpretation was considered in a dispute between Australia and France over French exports of wheat and flour; see BISD 1959, supplement 7, p. 46.
36. BISD 1980, supplement 26, p. 56 (69).

authoritative interpretation has not been shown to be sufficient to solve the problems of the interpretation of Article XVI(3); see immediately the following.

Over the years, the Union has made wide use of export subsidies for various agricultural products, and the current rules on this are in Articles 162–170 of the Single CMO Regulation.[37] The Union's export subsidies have led to several conflicts with the other GATT contracting parties.[38]

In 1978, Australia complained about the Union's export subsidies for sugar.[39] However, in its report the Panel did not find that the Union had obtained more than its equitable share of the world market. The Panel also stated that even if it had found that the Union had obtained more than an equitable share of the world market for sugar, it would have been difficult to establish that there was a causal connection between the Union's increased market share and Australia's loss of exports.[40] The Panel concluded that the Union's export support had put downward pressure on the price of sugar on the world market.[41] This triggered political negotiations to change the Union's support arrangements, though without leading to significant changes.[42]

[B]   The WTO Agreement on Agriculture

To the extent that export subsidies for agricultural products are permitted pursuant to Article XVI of the GATT Agreement, the WTO Agreement on Agriculture imposes further restrictions on the payment of such subsidies. The rules of the Agreement on Agriculture on export subsidies differ from its rules on domestic support. While the rules on domestic support regulate in detail what kinds of support *need not be* reduced, the Agreement's rules on export subsidies regulate in detail what forms of export support *must be* reduced.[43]

Article 3(3) of the Agreement on Agriculture provides that, as a main rule, a WTO Member may only provide export subsidies listed in Article 9(1) in respect of the

---

37. For the rules on export refunds in the Single CMO Regulation, see Ch. 4, §4.06[B][2].
38. It has been argued that the Union's export refunds were accepted in the GATT context in connection with the Dillon Round; see Neville-Rolfe, *The Politics of Agriculture in the European Community* 26 (Policy Studies Institute, 1984).
39. Brazil made the same complaint, and in this case the Panel gave a report that was in line with the Australian case; for the Brazil report see BISD 1981, supplement 27, 69.
40. BISD 1980, supplement 26, 290 (319, point f).
41. *Ibid.*, p. 319, point g. McMahon, *EU Agricultural Law* 76 (Oxford University Press, 2007), states that the Panel report 'may have been politically acceptable to the EC as it maintained the integrity of the system whilst forcing it to engage in consultations on possible ways of limiting the adverse effects of the system'. See also Usher, *EC Agricultural Law* 75 (Oxford University Press, 2d ed., 2001).
42. On the subsequent negotiations, see McMahon, *EU Agricultural Law* 76 (Oxford University Press, 2007); and Bentil, 'Attempts to Liberalize International Trade in Agriculture and the Problem of the External Aspects of the Common Agricultural Policy of the European Economic Community', *Case W. Res. J. Int'lL.* 1985, p. 335 (373).
43. See Usher, *EC Agricultural Law* 76 (Oxford University Press, 2d ed., 2001). For a comprehensive account of the rules on export support in the WTO Agreement on Agriculture and the associated practice, see McMahon, *EU Agricultural Law* 140–171 (Oxford University Press, 2007).

agricultural products listed in its Schedule, and only within specified limits.[44] Members may not pay export subsidies for agricultural products that are not listed in their Schedule.

However, the lawful forms of export support defined in Article 9(1) that are subject to reduction commitments cover not only publicly-funded export refunds, but also sales from public intervention stocks for export, and refunds for sales for exports of processed agricultural products.

Articles 10 and 11 of the WTO Agreement on Agriculture contain special rules to prevent the circumvention of the commitments to reduce export subsidies.

§9.06   THE WTO RULES IN UNION LAW

[A]   The Union's Treaties with Third Countries and International Organizations: The General Presumption

Treaties which the Union has entered into with third countries and other international organizations can according to Union law have direct effect in national law. This direct effect means that individual rules in such treaties in themselves create rights and duties for citizens or undertakings in the individual Member States. The rules in a treaty which the Union enters into with a third country or international organization that have direct effect must be applied in the form in which they appear in the treaty, and applied by the administrative authorities and courts of the Member States.[45] This effect of treaty provisions is not expressed in the TFEU, but has been established in the case law of the Court of Justice.

The *Kupferberg* case concerned a rule in a treaty in the form of a free-trade agreement between the European Community and Portugal, before Portugal became a Member State.[46] A provision of German law imposed a higher tax on port than on corresponding fortified wines produced in Germany. Kupferberg, a German wine importer, brought a case against the German tax and customs authorities before the German courts, arguing that the German tax provision was inapplicable because it was contrary to a provision in the free-trade agreement. This raised the question of whether the importer could rely on the free-trade agreement before the German courts, in other words whether the free-trade agreement had direct effect. This question was referred to the Court of Justice for a preliminary ruling. The Court of Justice established the general presumption that, depending on the content and structure of the treaty in question, the Union's treaties with third countries and other international organizations can have direct effect. The fact that a treaty is not given direct effect by the other contracting party or parties (third countries or other international organization) is not relevant in this context.[47]

---

44. There are exceptions to this main rule in Art. 3(3) and Art. 9(2)(b) and (4).
45. There is no need to distinguish between the terms 'direct effect' and 'directly applicable' either in theory or in practice.
46. Case 104/81 *Kupferberg* [1982] ECR 3641.
47. *Ibid.*, para. 18.

[B]   GATT and WTO Rules in Union Law

The general presumption about the effect of the Union's treaties with third countries and other international organizations in national law, as discussed in §9.06[A], does not apply to the rules of GATT or the WTO.

Prior to the setting up of the WTO, the Court of Justice ruled in the *International Fruit Company* case that the GATT Agreement, in its then current form, did not have direct effect in national law. The case concerned the import into the Union of dessert apples. Four Dutch fruit importers, including the International Fruit Company, sought licenses to import dessert apples from third countries, in accordance with the market organization for fruit and vegetables. When these applications were refused, the applicants brought proceedings before a national court and claimed that the Union legislative acts which formed the basis for the refusals were invalid because they were in conflict with Article XI of the GATT Agreement. The case was referred to the Court of Justice for a preliminary ruling. The Court of Justice ruled that the importers could not rely on Article XI before national courts.[48]

This result can be explained by the fact that it is important for the parties to the GATT Agreement to be able to take unilateral measures against other parties which breach the rules of the Agreement, and the Union's Member States would not have this possibility if the GATT Agreement had direct effect in national law.[49]

The setting up of the WTO did not change the practice of the Court of Justice. In an annulment case between Portugal and the Council, the Court of Justice emphasized the importance of the Union authorities' possibility of taking independent steps against parties that breach the WTO's rules. The Court stated that if it were accepted that the WTO rules had direct effect in the Union, this would 'deprive the legislative or executive organs of the [Union] of the scope for manoeuvre enjoyed by their counterparts in the [Union's] trading partners' which do not recognize the direct effect of the WTO rules.[50]

The position adopted in the annulment case has been maintained in subsequent cases, including in the *Fruchthandelgesellschaft* case. In this case, the German customs authorities collected customs duty from a German banana importer. The customs duty was collected on behalf of the Union under the authority of a regulation which was part of the market organization for bananas. Before the German court, the importer disputed the legality of the customs duty by reference to the fact that a WTO Panel had found that the market organization was contrary to Articles 1 and XIII of the GATT 1994 Agreement.

In its ruling, the Court of Justice stated that the provision in the relevant provisions in the GATT Agreement 'are not such as to create rights which individuals may rely on directly before a national court in order to oppose the application' of the market organization for bananas.[51]

---

48.   Joined Cases 21 to 24/72 *International Fruit Company NV and others* [1972] ECR 1219.
49.   See the Opinion of Advocate General Reischl in Joined Cases 267/81, 268/81 and 269/81 *SPI* [1983] ECR 801.
50.   Case C-149/96 *Portugal v. Council* [1999] ECR I-8395, paras 44–46.
51.   Case C-307/99 *Fruchthandelsgesellschaft* [2001] ECR I-3159.

The cases in which the Court of Justice has ruled that the WTO rules do not have direct effect have been subjected to strong criticism.[52] However, even if the case law of the Court of Justice is difficult to reconcile with the effect which the Court has given to other treaties from a jurisprudential perspective, the practice is understandable in the light of trade policy.[53] Section 102a(1) of the US Uruguay Round Agreements Act, states: 'UNITED STATES LAW TO PREVAIL IN CONFLICT – No provision of any of the Uruguay Round Agreements, nor the application of any such provision to any person or circumstance, that is inconsistent with any law of the United States shall have effect.'[54]

52. See Griller, 'Judicial Enforceablility of WTO Law in the European Union – Annotation to Case C-149/96, Portugal V. Council', J. Intl. Econ. L. 441 (2000), (446–447, references to the literature, with a summary on 472).
53. Peers, in Gráinne de Búrca & Scott, *The EU and the WTO: Legal and Constitutional Issues*, 111 (Hart, 2003) (summarized at 130); and Hartley, *The Foundations of European Union Law* 242 (Oxford University Press, 7th ed., 2010).
54. The US law is referred to by Usher, *EC Agricultural Law* 79 (Oxford University Press, 2d ed., 2001), 'to put the matter in context'. See Uruguay Round Agreements Act, Public Law, No 103-465, 103rd Congress, 2nd Session, 108 Statute 4809.

Bibliography

Ahner, Dirk. 'Europäisches Agrarrecht: Umsetzung der Reform – Ausblick in die Zukunft'. AUR, No. 8 (2005): Annex: Europäisches Agrarecht, 3–6.

Barents, René. 'The System of Deposits in Community Agricultural Law: Efficiency v. Proportionality'. ELR (1985): 239–249.

Barents, René. *The Agricultural Law of the EC*. Deventer, 1994.

Barents, René. 'Recent Developments in Community Case Law in the Field of Agriculture'. CMLRev (1997): 811–843.

Bellamy, Christopher & Graham Child. *European Community Law of Competition*, 5th edn, edited by Peter Roth. Oxford, 2001.

Bellamy, Christopher & Graham Child. *European Community Law of Competition*, 6th edn, edited by Peter Roth & Vivien Rose. Oxford, 2008.

Bentil. J. Kodwo. 'Attempts to Liberalize International Trade in Agriculture and the Problem of the External Aspects of the Common Agricultural Policy of the European Economic Community'. *Case W. Res. J. Int'l L.* (1985): 335–387.

Bercero, Garcia. 'Trade Laws, GATT and the Management of Trade Disputes between the US and the EEC'. YEL (1985): 149–189.

Bianchi, Daniele. *La politique agricole commune (PAC)*. 2nd edn. Bruxelles, 2012.

Bodiguel, Luc. 'The New European Rural Development Regulation: Implementation in France'. Drake J. Agric.L (2008): 7–19.

Cardwell, Michael. 'General principles of community law and milk quotas', CMLRev (1992): 723–747

Cardwell, Michael. *The European Model of Agriculture*. Oxford, 2004.

Cardwell, Michael. 'Rural Development in the European Community: Charting a New Course?' Drake J. Agric.L (2008): 21–61.

de Cockborne, Jean-Éric 'Les Règles communautaire de concurrence applicables aux entreprises dans le domaine agricole'. RTDE (1988): 293.

Constantinides-Megret, C. *La politique agricole commune en question*. Paris, 1982.

Cooke, John. 'Competition Rules and Agricultural Products'. In *Agricultural Law for the European Union*, edited by Heusel, Wolfgang & Anthony M. Collins, Trier, 1999: 67–85.

Danielsen, Jens Hartig. *Parallelhandel og varernes frie bevægelighed*. København, 2005.

Deringer, Arved. *Das Wettbewerbsrecht der Europäischen Wirtschaftsgemeinschaft*. Düsseldorf, 1962–1967.

Deringer, Arved. *The Competition Law of the European Economic Community*. New York, 1968.

Desta, Melaku Geboye. *The Law of International Trade in Agricultural Products*. Haag, 2002.

Diebold Jr., William. *The End of the ITO*. New Jersey: Princeton, 1952.

El-Agraa, Ali Mohammed. *The Economics of the Common Market*. 4th edn. New York, 1994.

Evald, Jens. *Landbrugets retsforhold III – Landbrugsrelateret ret*. København, 2007.

Evans, John W. *The Kennedy Round in American Trade Policy. The Twilight of the GATT?*. Massachusetts: Cambridge, 1971.

Filipek, Jon G. 'Agriculture in a World of Comparative Advantage: The Prospects for Farm Trade Liberalization in the Uruguay Round of GATT Negotiations'. HILJ (1989): 123–170.

Flynn, James. 'Force Majeure Pleas in Proceedings Before the European Court'. ELR (1981): 102–114.

Gilsdorf, Peter. 'La force majeure dans le droit de la CEE a la lumiere de la jurisprudence de la Cour de justice', CDE (1982): 137–143.

Griller, Stefan. 'Judicial Enforceablility of WTO Law in the European Union – Annotation to Case C-149/96, Portugal V. Council', JIEL (2000): 441–472.

Grimm, Christian. *Agrarrecht*, 2nd edn. München, 2004.

Harris, Simon. *E.E.C. Trade Relations with the U.S.A. in Agricultural Products*. London, 1977.

Hartley, Trevor C. *The Foundations of European Union Law: an introduction to the constitutional and administrative law of the European Union*. 7th edn. Oxford, 2010.

Hill, Brian E. *The Common Agricultural Policy: Past, present and future*. London, 1984.

Kaiser, Gerhard. *Grundriß des Agrarwirtschaftsrecht der Europäischen Union*. Wien, 1996.

Leidwein, Alois. *Einführung in das europäisches Agrarrecht*. Wien, 1996.

Leidwein, Alois. *Europaïsches Agrarrecht*, 2nd edn. Wien, 2004.

Loyant, Bruno. 'La force majeure et l'organisation commune des marchés agricole'. RTDE (1980): 255–283.

MacMahon, Joseph A. *Law of the common agricultural policy*. London, 2000.

McMahon, Joseph A. 'The Common Agricultural Policy: From Quantity to Quality?'. NILQ (2002): 9–27.

McMahon, Joseph A. *EU Agricultural Law*. Oxford, 2007.

MacMaoláin, Caoimhín. *EU Food Law*. Oxford, 2007.

Marsh, John S. & Swanney, Pamela J. *Agriculture and the European Community*. London, 1980.

Mensching, Jürgen. 'Farming Cooperatives and Competition Rules'. In *Agricultural Law for the European Union*, edited by Heusel, Wolfgang & Anthony M Collins. Trier, 1999, 87–97.

Meyer-Bolte Catharina. *Agrarrechtliche Cross Compliance als Steuerungsinstrument im Europäischen Verwaltungsverbund*. Baden-Baden, 2007.

Mortensen, Peter. *Landbrugets retsforhold II – Landbrugsloven*. Helsingør, 2005.

Neville-Rolfe, Edmund. *The Politics of Agriculture in the Community*. London, 1984.

O'Connor, Bernard. 'The Current Practice on Import Quotas: Problems and Solutions for Tariff Rate Quotas'. In *Agricultural Law for the European Union*, edited by Heusel, Wolfgang & Anthony M. Collins. Trier, 1999, 295–326.

Olmi, Gianscarlo. 'Common Organisation of Agricultural Markets at the Stage of the Single Market'. CMLRev (1967–1968): 359–408

Olmi, Giancarlo. *Le droit de la CEE (Commentaire Megret)*. vol. 2, 2nd edn. Politique agricole commune, 1991.

O'Reilly, James. 'Milk Quotas and their Consideration Before the Institutions of the Community of Particular Interest to Lawyers'. In *Agricultural Law for the European Union*, edited by Heusel, Wolfgang & Anthony M Collins. Trier, 1999, 101–111.

Ottervanger, Tom R. 'Antitrust and agriculture in the common market', FCLI (1990): 203–223.

Peers, Steve. 'Fundamental Right or Political Whim? WTO Law and the European Court of Justice'. In *The EU and the WTO: Legal and Constitutional Issues*, edited by Gráinne de Búrca & Joanne Scott. 2001, 111–130.

Picod, Fabrice. 'Libre circulation des produits agricoles et organisations communes de marchés'. In Heusel, *Agricultural Law for the European Union*, edited by Wolfgang & Anthony M. Collins Trier, 1999, 223–240.

Prieβ, Hans-Joachim. 'Practical Aspects og Obstacles to Trade and State Liability for Obstacles of a Private Nature'. In *Agricultural Law for the European Union*, edited by Heusel, Wolfgang & Anthony M Collins. Trier, 1999, 241–255

Le Roy, Pierre. *La Politique Agricole Commune*. Paris, 1994.

Ratliff, John. 'EC Competition Law and Agriculture'. In *Agricultural Law for the European Union*, edited by Heusel, Wolfgang & Anthony M. Collins. Trier, 1999, 37–59.

Ries, Adrien & Guida, Rosa Maria. 'L'application des règles de concurrence du traité CE à l'agriculture', CDE (1968): 60 and 165.

Schröter, Helmuth. Groeben, von Boeckh, Thiesing and Ehlermann, *Kommentar zum EWG-Vertrag*. 3rd edn. 1983.

Schulze, Frank. *Ratgeber Ararreform*. 2nd ed. Münster-Hiltrup, 1998.

Snyder, Francis G. *Law of the common agricultural policy*. London, 1985.

Sośniak, Mieczyslaw. 'Staatsakt und höhere Gewalt im internationalen Handelsverkehr der RGW-Länder'. RIW (1984): 105–108.

Thiele, Gereon. 'Weiterentwicklung der Rechtsprechung des Gerichtshofs der Europäischen Gemeinschaften (EuGH) zum Grundsatz des Vertrauensschutzes'. Agrarrecht (1988): 333–341.

Tiedemann, Paul. 'Rechtsprobleme der Agrarmarktintervention'. RIW (1980): 219–243.

Usher, John A. 'The Effects of Common Organisations and Policies on the Powers of a Member State'. ELR (1977): 428–443.

Usher, John A. 'Uniform External Protection - EEC Customs Legislation before the Court of Justice'. CMLRev (1982): 389–412.

Usher, John A. *EC Agricultural Law*. 2nd edn. Oxford, 2001.

Van Bael, Ivo & Bellis, Jean-François. *Competition Law of the European Community*. 4th edn. The Hague, 2005.

Warley, T.K. 'Western Trade in Agricultural Products'. In *International Economic Relations of the Western World*, edited by A. Schonfield. 1959–1971, Oxford, 1976, 285–402.

Whish, Richard. *Competition Law*. 7th edn. Oxford, 2011.

Wulff, Helge. *EU-jordbrugsret*. Frederiksberg, 2001.

Table of Cases

The Court of Justice and the General Court

Commissions Decisions

Unpublished judgment

Index

EUROPEAN MONOGRAPH SERIES

1. Lammy Betten (ed.), *The Future of European Social Policy*, 1991 (ISBN 90-654-4585-4).
2. Annemarie Loman, Kamiel Mortelmans, Harry H.G. Post & Stewart Watson, *Culture and Common Law: Before and After Maastricht*, 1992 (ISBN 90-654-4638-9).
3. John A.E. Vervaele, *Fraud against the Community: The Need for European Fraud Legislation*, 1994 (ISBN 90-654-4634-6).
4. Philip Raworth, *The Legislative Process in the European Community*, 1993 (ISBN 90-654-4690-7).
5. Jules Stuyck, *Financial and Monetary Integration in the European Economic Community*, 1993 (ISBN 90-654-4718-0).
6. Jules Stuyck & A.J. Vossestein (eds), *State Entrepreneurship, National Monopolies and European Community Law*, 1993 (ISBN 90-654-4773-3).
7. Jules Stuyck & A. Looijestijn-Clearie (eds), *The European Economic Area EC-EFTA*, 1994 (ISBN 90-654-4815-2).
8. Rosita B. Bouterse, *Competition and Integration: What Goals Count?*, 1995 (ISBN 90-654-4816-0).
9. René Barents, *The Agricultural Law of the EC: An Inquiry into the Administrative Law*, 1994 (ISBN 90-654-4867-5).
10. Nicholas Emiliou, *Principles of Proportionality in European Law: A Comparative Study*, 1996 (ISBN 90-411-0866-1).
11. Eivind Smith, *National Parliaments as Cornerstones of European Integration*, 1996 (ISBN 90-411-0898-X).
12. Jan H. Jans, *European Environmental Law*, 1996 (ISBN 90-411-0877-7).
13. Síofra O'Leary, *The Evolving Concept of Community Citizenship: From the Free Movement of Persons to Union Citizenship*, 1997 (ISBN 90-411-0878-5).
14. Laurence W. Gormley (ed.), *Current and Future Perspectives on EC Competition Law*, 1983 (ISBN 90-411-0691-X).
15. Simone White, *Protection of the Financial Interests of the European Communities: The Fight against Fraud and Corruption*, 1998 (ISBN 90-411-9647-1).
16. Morten P. Broberg, *Broberg on the European Commission's Jurisdiction to Scrutinise Mergers*, 4th Edition, 2013 (ISBN 978-90-411-3339-7).
17. Doris Hildebrand, *The Role of Economic Analysis in the EC Competition Rules: The European School*, 2nd Edition, 2002 (ISBN 90-411-1706-7).
18. Christof R.A. Swaak, *European Community Law and the Automobile Industry*, 1999 (ISBN 90-411-1140-9).

19. Dorthe Dahlgaard Dingel, *Public Procurement: A Harmonization of the National Judicial Review of the Application of European Community Law*, 1999 (ISBN 90-411-1161-1).
20. John A.E. Vervaele (ed.), *Compliance and Enforcement of European Community Law*, 1999 (ISBN 90-411-1151-4).
21. Martin Trybus, *European Defence Procurement Law: International and National Procurement Systems as Models for a Liberalized Defence Procurement Market in Europe*, 1999 (ISBN 90-411-1167-0).
22. Helen Staples, *The Legal Status of Third Country Nationals Resident in the European Union*, 1999 (ISBN 90-411-1277-4).
23. Damien Geradin (ed.), *The Liberalization of State Monopolies in the European Union and Beyond*, 1999 (ISBN 90-411-1264-2).
24. Katja Heede, *European Ombudsman: Redress and Control at Union Level*, 2000 (ISBN 90-411-1413-0).
25. Ulf Bernitz & Joakim Nergelius (eds), *General Principles of European Community Law*, 2000 (ISBN 90-411-1402-5).
26. Michaela Drahos, *Convergence of Competition Laws and Policies in the European Community*, 2002 (ISBN 90-411-1562-5).
27. Damien Geradin (ed.), *The Liberalization of Electricity and Natural Gas in the European Union*, 2001 (ISBN 90-411-1560-9).
28. Gisella Gori, *Towards an EU Right to Education*, 2001 (ISBN 90-411-1670-2).
29. Brendan P.G. Smith, *Constitution Building in the European Union*, 2001 (ISBN 90-411-1695-8).
30. Friedl Weiss & Frank Wooldridge, *Free Movement of Persons within the European Community*, 2nd Edition, 2007 (ISBN 978-90-411-2545-3).
31. Ingrid Boccardi, *Europe and Refugees: Towards an EU Asylum Policy*, 2002 (ISBN 90-411-1709-1).
32. John A.E. Vervaele & André Klip (eds), *European Cooperation Between Tax, Customs and Judicial Authorities*, 2001 (ISBN 90-411-1747-4).
33. Wouter P.J. Wils, *The Optimal Enforcement of EC Antitrust Law: Essays in Law and Economics*, 2002 (ISBN 90-411-1757-1).
34. Damien Geradin (ed.), *The Liberalization of Postal Services in the European Union*, 2002 (ISBN 90-411-1780-6).
35. Nick Bernard, *Multilevel Governance in the European Union*, 2002 (ISBN 90-411-1812-8).
36. Jill Wakefield, *Judicial Protection through the Use of Article 288(2) EC*, 2002 (ISBN 90-411-1823-3).
37. Sebastiaan Princen, *EU Regulation and Transatlantic Trade*, 2002 (ISBN 90-411-1871-3).

57. Rass Holdgaard, *External Relations Law of the European Community: Legal Reasoning and Legal Discourses,* 2007 (ISBN 978-90-411-2604-7).
58. Jill Wakefield, *The Right to Good Administration,* 2007 (ISBN 978-90-411-2697-9).
59. Dimitry Kochenov, *EU Enlargement and the Failure of Conditionality: Pre- accession Conditionality in the Fields of Democracy and the Rule of Law,* 2008 (ISBN 978-90-411-2696-2).
60. Despina Mavromati, *The Law of Payment Services in the EU: The EC Directive on Payment Services in the Internal Market,* 2008 (ISBN 978-90-411-2700-6).
61. Anne Meuwese, *Impact Assessment in EU Lawmaking,* 2008 (ISBN 978-90-411-2720-4).
62. Ulf Bernitz, Joakim Nergelius & Cecilia Cardner (eds), *General Principles of EC Law in a Process of Development,* 2008 (ISBN 978-90-411-2705-1).
63. Johan van de Gronden (ed.), *The EU and WTO Law on Services: Limits to the Realisation of General Interest Policies within the Services Markets?,* 2008 (ISBN 978-90-411-2809-6).
64. Alina Tryfonidou, *Reverse Discrimination in EC Law,* 2009 (ISBN 978-90-411-2751-8).
65. Mikael Berglund, *Cross-Border Enforcement of Claims in the EU: History Present Time and Future,* 2009 (ISBN 978-90-411-2861-4).
66. Theodore Konstadinides, *Division of Powers in European Union Law: The Delimitation of Internal Competence between the EU and the Member States,* 2009 (ISBN 978-90-411-2615-3).
67. Mattias Derlén, *Multilingual Interpretation of European Union Law,* 2009 (ISBN 978-90-411-2853-9).
68. René Barents, *Directory of EU Case Law on the Preliminary Ruling Procedure,* 2009 (ISBN 978-90-411-3150-8).
69. Yan Luo, *Anti-dumping in the WTO, the EU and China: The Rise of Legalization in the Trade Regime and its Consequences,* 2010 (ISBN 978-90-411-3207-9).
70. Patrick Birkinshaw & Mike Varney (eds), *The European Union Legal Order after Lisbon,* 2010 (ISBN 978-90-411-3152-2).
71. Thomas Gr. Papadopoulos, *EU Law and Harmonization of Takeovers in the Internal Market,* 2010 (978-90-411-3340-3).
72. Bas van Bockel, *The* Ne Bis In Idem *Principle in EU Law,* 2010 (978-90-411-3156-0).
73. Veljko Milutinović, *The 'Right to Damages' under EU Competition Law: From* Courage v. Crehan *to the White Paper and Beyond,* 2010 (978-90-411-3235-2).

74. Amandine Garde, *EU Law and Obesity Prevention*, 2010 (978-90-411-2706-8).
75. Leonard Besselink, Frans Pennings & Sacha Prechal (eds), *The Eclipse of the Legality Principle in the European Union*, 2011 (978-90-411-3262-8).
76. Sacha Garben, *EU Higher Education Law: The Bologna Process and Harmonization by Stealth*, 2011 (978-90-411-3365-6).
77. Dimitry Kochenov (ed.), *EU Law of the Overseas: Outermost Regions, Associated Overseas Countries and Territories, Territories Sui Generis*, 2011 (978-90-411-3445-5).
78. Pablo Ibáñez Colomo, *European Communications Law and Technological Convergence: Deregulation, Re-regulation and Regulatory Convergence in Television and Telecommunications*, 2012 (978-90-411-3829-3).
79. Elise Muir, *EU Regulation of Access to Labour Markets: A Case Study of EU Constraints on Member State Competences*, 2012 (978-90-411-3823-1).
80. Tim Corthaut, *EU Ordre Public*, 2012 (978-90-411-3232-1).
81. Oana Ştefan, *Soft Law in Court: Competition Law, State Aid and the Court of Justice of the European Union*, 2013 (978-90-411-3997-9).
82. Francesco Rossi dal Pozzo, *Citizenship Rights and Freedom of Movement in the European Union*, 2013 (978-90-411-4660-1).
83. Jens Hartig Danielsen, *EU Agricultural Law*, 2013 (978-90-411-3280-2).